KEY TECHNOLOGIES IN
POLYMER CHEMISTRY

AAP Research Notes on
Polymer Engineering Science and Technology

KEY TECHNOLOGIES IN POLYMER CHEMISTRY

Edited by
Nikolay D. Morozkin, DSc,
Vadim P. Zakharov, DSc, and
Gennady E. Zaikov, DSc

APPLE
ACADEMIC
PRESS

Apple Academic Press Inc. | Apple Academic Press Inc.
3333 Mistwell Crescent | 9 Spinnaker Way
Oakville, ON L6L 0A2 | Waretown, NJ 08758
Canada | USA

©2015 by Apple Academic Press, Inc.

First issued in paperback 2021

Exclusive worldwide distribution by CRC Press, a member of Taylor & Francis Group

No claim to original U.S. Government works

ISBN 13: 978-1-77463-343-4 (pbk)
ISBN 13: 978-1-77188-024-4 (hbk)

Library and Archives Canada Cataloguing in Publication

Key technologies in polymer chemistry / edited by Nikolay D. Morozkin, DSc, Vadim P. Zakharov, DSc, and Gennady E. Zaikov, DSc.

(AAP research notes on polymer engineering science and technology series)
Includes bibliographical references and index.
ISBN 978-1-77188-024-4 (bound)
1. Polymerization. I. Zaikov, G. E. (Gennadiĭ Efremovich), 1935- author, editor II. Morozkin, Nikolay D., author, editor III. Zakharov, Vadim P. (Vadim Petrovich), author, editor IV. Series: AAP research notes on polymer engineering science and technology series

QD281.P6K49 2015   547'.28   C2014-907858-7

Library of Congress Cataloging-in-Publication Data

Key technologies in polymer chemistry / Nikolay D. Morozkin, DSc., Vadim P. Zakharov, DSc., and Gennady E. Zaikov, DSc., editors.

pages cm. -- (AAP research notes on polymer engineering science and technology)
Results of research by Bashkir State University, Ufa, Russia.
Includes bibliographical references and index.
ISBN 78-1-77188-024-4 (alk. paper)
1. Polymers. 2. Supramolecular chemistry. 3. Chemistry. I. Morozkin, Nikolay D., editor. II. Zakharov, Vadim P. (Vadim Petrovich), editor. III. Zaikov, G. E. (Gennadii Efremovich), 1935- editor. IV. Bashqort daulat universitety.

QD381.K44 2015   668.9--dc23   2014045762

Apple Academic Press also publishes its books in a variety of electronic formats. Some content that appears in print may not be available in electronic format. For information about Apple Academic Press products, visit our website at **www.appleacademicpress.com** and the CRC Press website at **www.crcpress.com**

AAP Research Notes on
Polymer Engineering Science and Technology

KEY TECHNOLOGIES IN POLYMER CHEMISTRY

Edited by
**Nikolay D. Morozkin, DSc,
Vadim P. Zakharov, DSc, and
Gennady E. Zaikov, DSc**

Apple Academic Press Inc. | Apple Academic Press Inc.
3333 Mistwell Crescent | 9 Spinnaker Way
Oakville, ON L6L 0A2 | Waretown, NJ 08758
Canada | USA

©2015 by Apple Academic Press, Inc.

First issued in paperback 2021

Exclusive worldwide distribution by CRC Press, a member of Taylor & Francis Group

No claim to original U.S. Government works

ISBN 13: 978-1-77463-343-4 (pbk)
ISBN 13: 978-1-77188-024-4 (hbk)

Library and Archives Canada Cataloguing in Publication

Key technologies in polymer chemistry / edited by Nikolay D. Morozkin, DSc, Vadim P. Zakharov, DSc, and Gennady E. Zaikov, DSc.

(AAP research notes on polymer engineering science and technology series)
Includes bibliographical references and index.
ISBN 978-1-77188-024-4 (bound)
1. Polymerization. I. Zaikov, G. E. (Gennadiĭ Efremovich), 1935- author, editor II. Moro-zkin, Nikolay D., author, editor III. Zakharov, Vadim P. (Vadim Petrovich), author, editor IV. Series: AAP research notes on polymer engineering science and technology series

QD281.P6K49 2015   547'.28   C2014-907858-7

Library of Congress Cataloging-in-Publication Data

Key technologies in polymer chemistry / Nikolay D. Morozkin, DSc., Vadim P. Zakharov, DSc., and Gennady E. Zaikov, DSc., editors.

pages cm. -- (AAP research notes on polymer engineering science and technology)
Results of research by Bashkir State University, Ufa, Russia.
Includes bibliographical references and index.
ISBN 78-1-77188-024-4 (alk. paper)
1. Polymers. 2. Supramolecular chemistry. 3. Chemistry. I. Morozkin, Nikolay D., editor. II. Zakharov, Vadim P. (Vadim Petrovich), editor. III. Zaikov, G. E. (Gennadii Efremovich), 1935- editor. IV. Bashqort daulat universitety.

QD381.K44 2015   668.9--dc23   2014045762

Apple Academic Press also publishes its books in a variety of electronic formats. Some content that appears in print may not be available in electronic format. For information about Apple Academic Press products, visit our website at **www.appleacademicpress.com** and the CRC Press website at **www.crcpress.com**

ABOUT THE EDITORS

Nikolay D. Morozkin, DSc

Nikolay D. Morozkin, DSc, is a Professor and Rector at Bashkir State University in Ufa, Russia. He is an Honored Worker of Russian Higher Education and an Honored Scientific Worker of the Republic of Bashkortostan. He has trained eight candidates of science (PhD) and is author and co-author of more than 100 scientific works. His area of expertise is applied mathematics and informatics.

Vadim P. Zakharov, DSc

Vadim P. Zakharov, DSc, is Professor and Vice-Chancellor for Scientific Work at Bashkir State University in Ufa, Russia. He is a Laureate of the Russian Government Award in Science and Technology. He has trained five candidates of science (PhD) and is author and co-author of three monographs, more than 160 scientific articles, and 11 patents. His area of expertise is in macrokinetics of high-speed processes in liquid phase and catalytic polymerization of dienes.

Gennady E. Zaikov, DSc

Gennady E. Zaikov, DSc, is Head of the Polymer Division at the N. M. Emanuel Institute of Biochemical Physics, Russian Academy of Sciences, Moscow, Russia, and professor at Moscow State Academy of Fine Chemical Technology, Russia, as well as professor at Kazan National Research Technological University, Kazan, Russia. He is also a prolific author, researcher, and lecturer. He has received several awards for his work, including the the Russian Federation Scholarship for Outstanding Scientists. He has been a member of many professional organizations and on the editorial boards of many international science journals.

ABOUT AAP RESEARCH NOTES ON POLYMER ENGINEERING SCIENCE AND TECHNOLOGY

The AAP Research Notes on Polymer Engineering Science and Technology reports on research development in different fields for academic institutes and industrial sectors interested in polymer engineering science and technology. The main objective of this series is to report research progress in this rapidly growing field.

Editor-in-Chief: Sabu Thomas, PhD
Director, School of Chemical Sciences, Professor of Polymer Science & Technology & Honorary Director of the International and Inter University Centre for Nanoscience and Nanotechnology, Mahatma Gandhi University, Kottayam, India
email: sabupolymer@yahoo.com

BOOKS IN THE AAP RESEARCH NOTES ON POLYMER ENGINEERING SCIENCE AND TECHNOLOGY SERIES

CONTENTS

LIST OF CONTRIBUTORS

M. I. Abdullin
Bashkir State University, ZakiValidiStr. 32, Ufa, 450076, Bashkortostan, Russia

I. Sh. Ahatov
Bashkir State University, ZakiValidiStr. 32, Ufa, 450076, Bashkortostan, Russia

R. M. Akhmetkhanov
Bashkir State University, ZakiValidiStr. 32, Ufa, 450076, Bashkortostan, Russia

G. R. Akmanova
Bashkir State University, ZakiValidiStr. 32, Ufa, 450076, Russia

A.V. Allagulova
Bashkir State University, ZakiValidiStr. 32, Ufa, 450076, Bashkortostan, Russia

A. D. Badikova
Bashkir State University, ZakiValidiStr. 32, Ufa, 450076, Bashkortostan, Russia

A. A. Basyrov
Bashkir State University, ZakiValidiStr. 32, Ufa, 450076, Bashkortostan, Russia

M. V. Bazunova
Bashkir State University, ZakiValidiStr. 32, Ufa, 450076, Bashkortostan, Russia

Z. A. Berezhneva
Bashkir State University, ZakiValidiStr. 32, Ufa, 450076, Bashkortostan, Russia

Yu. N. Biglova
Bashkir State University, ZakiValidiStr. 32, Ufa, 450076, Bashkortostan, Russia

N. N. Bikkulova
Sterlitamak branch of Bashkir State University, 37 Lenin Avenue, Sterlitamak 453103, Russia

D. M. Bikmeev
Bashkir State University, ZakiValidiStr. 32, Ufa, 450076, Bashkortostan, Russia

I. M. Borisov
Bashkir State University, ZakiValidiStr. 32, Ufa, 450076, Russia

V. V. Chernova
Bashkir State University, ZakiValidiStr. 32, Ufa, 450076, Bashkortostan, Russia

A. D. Davletshina
Bashkir State University, ZakiValidiStr. 32, Ufa, 450076, Russia

E. G. Ekomasov
Bashkir State University, ZakiValidiStr. 32, Ufa, 450076, Bashkortostan, Russia

R. R. Faizullina
Bashkir State University, ZakiValidiStr. 32, Ufa, 450076, Bashkortostan, Russia

A. Ya. Gerchikov
Bashkir State University, ul. ZakiValidy 32, Ufa, Bashkortostan, 450076, Russia

R. N. Gimaev
Bashkir State University, ZakiValidiStr. 32, Ufa, 450076, Bashkortostan, Russia

A. B. Glazyrin
Bashkir State University, ZakiValidiStr. 32, Ufa, 450076, Bashkortostan, Russia

A. V. Grigoryeva
Bashkir State University, ZakiValidiStr. 32, Ufa, 450076, Bashkortostan, Russia

A. M. Gumerov
Bashkir State University, ZakiValidiStr. 32, Ufa, 450076, Bashkortostan, Russia

R. I. Ibragimov
Bashkir State University, ZakiValidiStr. 32, Ufa, 450076, Bashkortostan, Russia

I. M. Kamaltdinov
Bashkir State University, ZakiValidiStr. 32, Ufa, 450076, Bashkortostan, Russia

V. R. Khairullina
Bashkir State University, ul. ZakiValidy 32, Ufa, Bashkortostan, 450076, Russia

S. V. Kolesov
Bashkir State University, ZakiValidiStr. 32, Ufa, 450076, Bashkortostan, Russia

F. Kh. Kudasheva
Bashkir State University, ZakiValidiStr. 32, Ufa, 450076, Bashkortostan, Russia

O. S. Kukovinets
Bashkir State University, ZakiValidiStr. 32, Ufa, 450076, Bashkortostan, Russia

E. I. Kulish
Bashkir State University, ZakiValidiStr. 32, Ufa, 450076, Bashkortostan, Russia

O. Yu. Kupova
Bashkir State University, ZakiValidiStr. 32, Ufa, 450076, Bashkortostan, Russia

M. V. Mavletov
Bashkir State University, ZakiValidiStr. 32, Ufa, 450076, Bashkortostan, Russia

V. N. Maystrenko
Bashkir State University, ZakiValidiStr. 32, Ufa, 450076, Bashkortostan, Russia

V. Z. Mingaleev
Institute of Organic Chemistry, Ufa Scientific Center of Russian Academy of Sciences, pr. Oktyabrya 71, Ufa, Bashkortostan, 450054, Russia

L. A. Murzina
Bashkir State University, ZakiValidiStr. 32, Ufa, 450076, Bashkortostan, Russia

A. G. Mustafin

Bashkir State University, ZakiValidiStr. 32, Ufa, 450076, Bashkortostan, Russia

T. V. Sharipov

Bashkir State University, ZakiValidiStr. 32, Ufa, 450076, Bashkortostan, Russia

F. B. Shevlyakov

Ufa State Petroleum Technical University, 1Kosmonavtov St. Ufa, 45006, Russia

I. A. Shpirnaya

Bashkir State University, ZakiValidiStr. 32, Ufa, 450076, Bashkortostan, Russia

A. S. Shurshina

Bashkir State University, ZakiValidiStr. 32, Ufa, 450076, Bashkortostan, Russia

A. V. Sidelnikov

Bashkir State University, ZakiValidiStr. 32, Ufa, 450076, Bashkortostan, Russia

R. R. Syrlybaeva

Bashkir State University, ZakiValidiStr. 32, Ufa, 450076, Bashkortostan, Russia

I. A. Taipov

Institute of Organic Chemistry, Ufa Scientific Center of Russian Academy of Sciences, pr. Oktyabrya 71, Ufa, Bashkortostan, 450054, Russia

R. F. Talipov

Bashkir State University, ZakiValidiStr. 32, Ufa, 450076, Bashkortostan, Russia

G. R. Talipova

Bashkir State University, ZakiValidiStr. 32, Ufa, 450076, Bashkortostan, Russia

V. O. Tsvetkov

Bashkir State University, ZakiValidiStr. 32, Ufa, 450076, Bashkortostan, Russia

I.F. Tuktarova

Bashkir State University, ZakiValidiStr. 32, Ufa, 450076, Bashkortostan, Russia

T. G. Umergalin

Ufa State Petroleum Technical University,1 Kosmonavtov St. Ufa, 45006, Russia

I. V. Vakulin

Bashkir State University, ZakiValidiStr 32, Ufa, 450076, Bashkortostan, Russia

E. R. Valinurova

Bashkir State University, ZakiValidiStr. 32, Ufa, 450076, Bashkortostan, Russia

M. N. Vasilyev

Bashkir Medical University, 3 Lenin Str., 450000, Ufa, Russia

R. A. Yakshibaev

Bashkir State University, ZakiValidiStr. 32, Ufa, 450076, Russia

V. P. Zakharov

Bashkir State University, ZakiValidiStr. 32, Ufa, 450076, Bashkortostan, Russia

E. M. Zakharova

Institute of Organic Chemistry, Ufa Scientific Center of Russian Academy of Sciences, pr. Oktyabrya 71, Ufa, Bashkortostan, 450054, Russia

I. D. Zakirova

Institute of Organic Chemistry, Ufa Scientific Center of Russian Academy of Sciences, pr. Oktyabrya 71, Ufa, Bashkortostan, 450054, Russia

F. S. Zarudiy

Institute of Organic Chemistry of Ufa Research Centre of the Russian Academy of Sciences, 71 Pros-pektOctyabrya, 450054, Ufa, Republic of Bashkortostan, Russia

F. S. Zarydiy

Bashkir State University, ul. ZakiValidy 32, Ufa, Bashkortostan, 450076, Russia

Yu. S. Zimin

Bashkir State University, ZakiValidiStr. 32, Ufa, 450076, Russia

LIST OF ABBREVIATIONS

AAA	Antiarrhythmic Activity
AAT	Antiarrhythmic Therapy
AFM	Atomic Force Microscope
AMS	Amikacin Sulfate
CMC	Carboxymethylcellulose
CPE	Carbon-Paste Electrodes
DSD	Deciding Set of Descriptor
EDP	Electric Desalting Plant
FC	Frozen Core
FD	Fragment Descriptors
FO	Formaldehyde Oligomers
FP-LMTO	Full-Potential Linear Muffin-Tin Orbital
GMS	Gentamicin Sulfate
GUSAR	General Unrestricted Structure Activity Relationships
LDA	Local Density Approximation
LMO	Leave Many Out
MNA	Multilevel Neighborhoods of Atoms
PCM	Polarizable Continuum Model
PVA	Polyvinyl Alcohol
QNA	Quantitative Neighborhoods of Atoms
SARD	Structure Activity Relationship and Design
SIMCA	Soft Independent Modeling of Class Analogy
SPR	Structure-Property Relationship

LIST OF SYMBOLS

a	effective parameters
A_{hi}	hydrodynamic invariant
C	sign (activity)
c_{ed}	concentration of chitosan
d	diameter
D	diffusion constants
F	rules of recognition
g	slope of the line
k	effective speed constant
k_2	decay constant of the enzyme
L	length
n_1	number of structures in the inactive compound
N_A	Avogadro number
N_{pair}	number of unshared electron pairs
N_{val}	number of valence electrons
N_{core}	number of core electrons
R	gas constant
S_0	sedimentation constant
S_c	extrapolation on the zero concentration
T	absolute temperature
t	time
v	partial specific gravity
w	speed
W	width of the spatial modulation
x	maximum coordinate
x_0	centre coordinate of the solution
x_1	kink
ρ	density
σ	tensile strength
γ	alkylsubstitutedcations
ζ	coefficient of local resistance
$\delta(x)$	dirac delta function

Δh	coordinate step
λ	friction coefficient
η_2	outputs of excitation and emission
α	proportions of peroxy acids
Δt	time step
η	intrinsic viscosity
η_0	dynamic viscosity
ν	partial specific volume
ρ_0	density of the solvent
ω	breather oscillation frequency

PREFACE

This book presents the results of scientific research carried out by the staff of Bashkir State University in the priority areas of chemistry, biology and physics.

In the field of polymer chemistry the research is done on the patterns of obtaining biodegradable polymer films from polyethylene, which is modified by the natural biopolymers under intensive shear deformation. It is presented that chemical modification of syndiotactic 1,2-polybutadienecan allows producing polymer compounds with new physical and mechanical properties. New methods of modifying natural polymer chitosan are proposed for the purpose of creating new medical film coatings to be used for surgical treatment of wounds and burns.

New methods of obtaining nanopatterned structures as well as their properties are introduced and described: multilayer particles of nanocarbon, methanol fullerene, modified hydrocarbon fibers, solid solutions of cuprum and argentum chalcogenides.

In the sphere of environmental issues the following subjects are considered: technological methods of waste disposal of concentrated sulfuric acid, analytical methods of motor oils identification with a use of carbon-based electronic language, as well as kinetics patterns of the oxidation of water-soluble organic compounds.

The work proposes methodological approaches to the development of new high-performance antiarrhythmic and antineoplastic drugs, as well as to the restructuring of existent drugs to enhance efficiency of their physiological effect on the human body.

In the sphere of mathematical and quantum chemistry the work examines regularities of synthesis of substituted 1,3-dioxins as a result of the Prins reaction.

The editors would much appreciate readers' comments and suggestions and will take them into consideration in future work.

— **Nikolay D. Morozkin, DSc,**
Vadim P. Zakharov, DSc, and
Gennady E. Zaikov, DSc

PART I

MATERIALS WITH IMPROVED PROPERTIES AT SYNTHESIS AND MODIFICATION OF POLYMERS

CHAPTER 1

PECULIARITIES OF METHANOFULLERENE STRUCTURE REVEALED BY UV-SPECTROSCOPY

YU. N. BIGLOVA

CONTENTS

ABSTRACT

The methanofullerene compounds differing in the functionalization degree were researched by UV-spectroscopy. It has been shown that the increase in the substituent number results in a hypsochromic shift of the characteristic absorption bands and reduction of the optic density of the solutions studied in the characteristic maxima. The same dependence is observed for optic densities characteristic for fullerene bands at 258 and 329 nm (A_{258}/A_{329}). The exponential dependence of methanofullerene molar absorption coefficients from the fullerene nucleus functionalization degree is found

1.1   INTRODUCTION

The usage of fullerenes as additives that change physical and mechanical, electric and physical and other polymers promises great perspectives [1, 2]. However, the fullerene is limitedly dissolved in organic solvents thus preventing its introduction into polymer compositions in considerable quantities. In this case the synthesis and study of the properties of highly soluble functionally substituted fullerenes are of great interest thus resulting in the increasing number of publications. A considerable part of these works is related to the synthesis and study of methanofullerene properties obtained under the Bingol Hirsch reaction, which results in cyclopropane adducts [3, 4]. It should be stated that as a rule C_{60} functionalization is accompanied by the formation of the mixture of hard-shared products of different degree of substitution.

UV spectroscopy may serve as one of the qualitative and quantitative ways of fullerene adducts analysis in such mixtures [5–7]. In literary sources there is no data on spectral characteristics of fullerene compounds with different degrees of substitution, which allows to estimate the content of the functionalization of products in their mixtures. This paper is dedicated to the UV spectroscopy study of C_{60} transformation into methanofullerene derivatives with different functionalization degrees.

1.2   EXRERIMENTAL PART

Fullerene C_{60} (JSC ILIP, 99.5% purity) was used without further purification. Methanofullerenes with different degrees of substitution,

[(1¢,1¢-diallyloxycarbonyl)-1,2-methane]-1,2-dihydro-C_{60}-fullerene (I); [(1¢,1¢-diethyloxycarbonyl)-1,2-methane]-1,2-dihydro-C_{60}-fullerene (II); [1-(methacryloyloxy)ethyloxycarbonyl-1-chlorine)methane]-1,2-dihydro-C_{60}-fullerene (III); {(1-methoxycarbonyl-1-[(methacryloyloxy)-ethyloxycarbonyl]-1,2-methane]}-1,2-dihydro-C_{60}-fullerene (IV); [(1-methoxycarbonyl)-1-(propynyloxycarbonyl)methane]-1,2-dihydro-C_{60}-fullerene (V); {(1-methoxycarbonyl-1-(10-undecenyloxycarbonyl)-1,2-methane]}-1,2-dihydro-C_{60}-fullerene (VI), were synthesized by the modified Bingel–Hirsch method [8–10]:

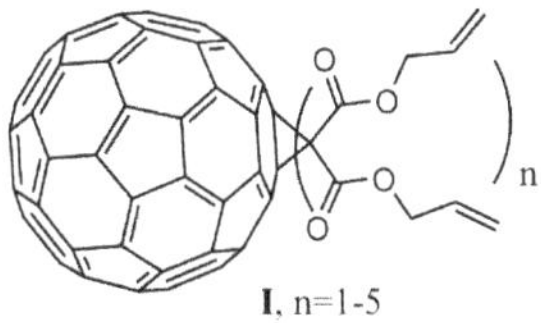

I, n=1-5

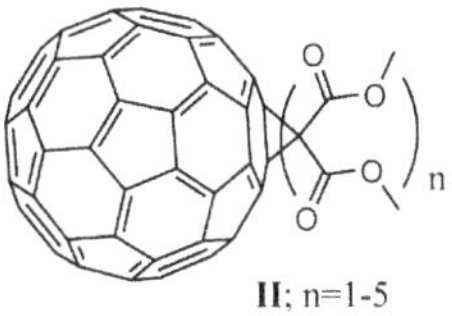

II; n=1-5

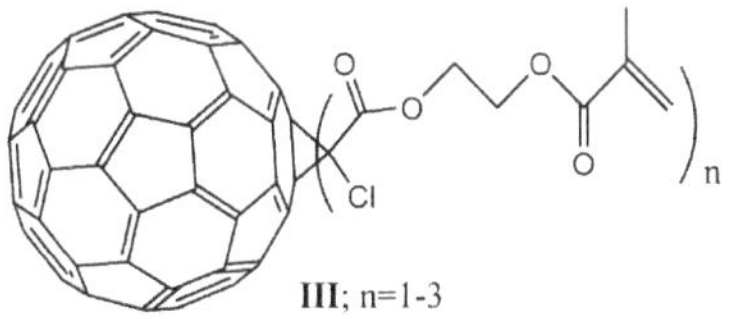

III; n=1-3

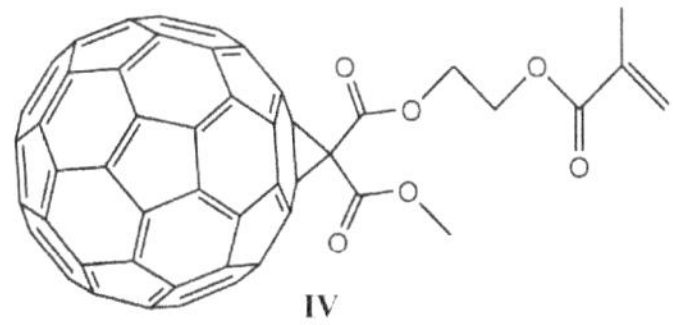

IV

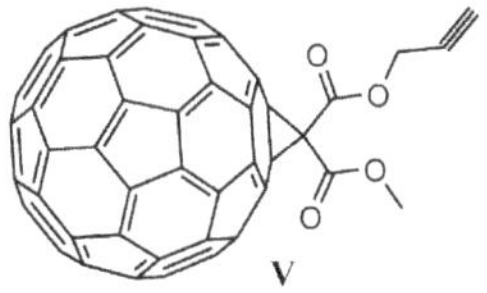

V

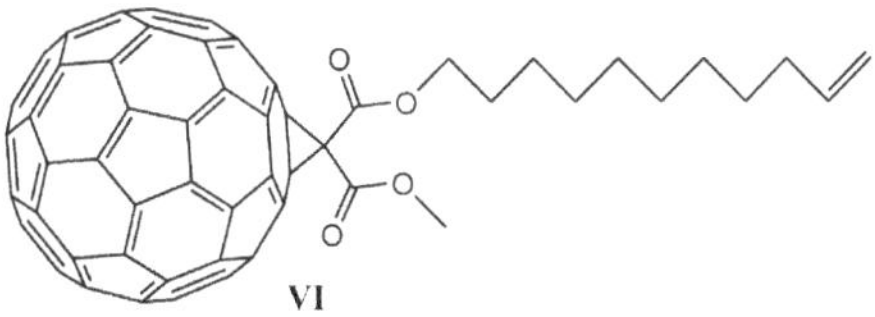

VI

Single peaks of molecular ions testifying the adduct purity were marked for the chosen regioisomeric substances in mass-spectra. The construction of concrete isomers was not established.

Acetonitrile (SP), chloroform (high grade) were used as solvents and were purified by standard methods. As a result all their physical and chemical constants corresponded to the reference data.

Electronic absorption spectra of the researched compounds were recorded in the coordinates of optical density (A), wavelength (λ) in the UV-1240 spectrophotometer UV-mini Sumadzu. A spectrum recording was carried out at room temperature, with wavelength ranging from 190 nm to

1100 nm (slit width 2.0 nm, scanning speed fast), using a quartz cuvette of 1 cm thick.

Molar absorption coefficients were determined from the concentration dependences of the optical solution density measured at the maxima of the absorption bands.

The adduct calibration solutions were prepared by a direct dissolution of methane derivatives. The concentration of the solutions ranged from 10^{-6} to 10^{-4} mol/L and aliquot parts were taken by micropipettes of Biohit Proline series (Finland). The sample weighing was carried on the Sartorius M2P scales (of 10^{-6} g sensitivity).

1.3   RESULTS AND DISCUSSION

1.3.1   SPECTRAL CHARACTERISTICS OF MONO-SUBSTITUTED METHANOFULLERENES

The spectra of a series of mono-substituted 1,2-dihydro-C_{60}-fullerenes registered in similar conditions (Fig. 1.1) revealed the following:

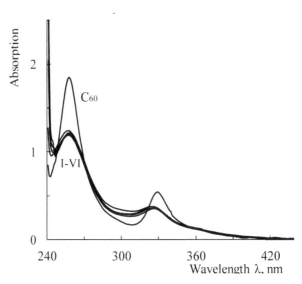

FIGURE 1.1   UV spectra of C_{60} and series of I-VI mono-substituted 1,2-dihydro-C_{60}-fullerenes taken in chloroform in the same 10^{-5} mol/L concentrations.

Regardless of the nature and length of the substituents, spectral curves of mono-substituted 1,2-dihydro-C_{60}-fullerenes are identical: the spectra of all modified fullerenes of a single concentration almost coincide (Fig. 1.1). While transiting from the initial C_{60} to its mono-substituted derivatives, the maxima of the absorption bands are shifted to short-wave regions whereas the absorption intensity is reduces by 35% approximately.

For all bands in the differential spectrum a linear dependence of the optical density and the concentration of the compounds is observed. The obtained results of the spectral characteristics of a number of mono-substituted 1,2-dihydro-C_{60}-fullerenes are shown in Table 1.1.

TABLE 1.1 Changes of Molar Absorption Coefficients Depending on Mono-Substituted 1,2-Dihydro-C_{60}-Fullerenes Concentration

Compound	Concentration $\times 10^5$, mol/L	$\varepsilon(327\ nm)$, $M^{-1} \times cm^{-1}$	$\varepsilon(258\ nm)$, $M^{-1} \times cm^{-1}$	A_{258}/A_{327}
I	0.46	34,600	125,890	3.64
	0.73	34,260	119,500	3.49
	1.09	35,110	121,510	3.46
	1.44	35,340	119,960	3.39
	2.14	36,350	118,040	3.25
Average extinction coefficient		35,720	118,630	$\varepsilon_{424}=1870$
		R=0.9984	R=0.9979	$M^{-1} \times cm^{-1}$
II	0.31	36,960	123,220	3.33
	0.61	37,000	122,800	3.32
	0.91	36,700	121,200	3.30
	1.20	36,730	119,780	3.26
	2.13	35,540	114,850	3.23
Average extinction coefficient		35,950	116,850	$\varepsilon_{424}=2040$
		R=0.9993	R=0.9985	$M^{-1} \times cm^{-1}$

III	0.50	38,000	127,800	3.36
	0.59	37,670	125,680	3.34
	1.13	37,690	124,190	3.30
	2.19	36,350	115,600	3.18
	2.60	36,090	110,200	3.05
Average extinction coefficient		36,060	113,020	$\varepsilon_{424}=1980$
		R=0.9988	R=0.9947	$M^{-1}\times cm^{-1}$
IV	0.49	36,250	121,790	3.36
	0.73	36,160	121,230	3.35
	0.83	36,140	119,520	3.31
	1.04	35,480	116,150	3.27
	1.26	35,320	115,000	3.25
Average extinction coefficient		35,660	117,260	$\varepsilon_{424}=2390$
		R=0.9995	R=0.9980	$M^{-1}\times cm^{-1}$
V	0.40	37,000	123,750	3.34
	0.49	36,740	122,250	3.33
	0.63	36,540	121,260	3.32
	1.54	35,840	117,660	3.28
	1.76	35,460	116,080	3.27
Average extinction coefficient		35,760	117,440	$\varepsilon_{424}=2150$
		R=0.9997	R=0.9993	$M^{-1}\times cm^{-1}$
VI	0.77	39,010	124,030	3.18
	0.96	37,760	120,250	3.18
	1.15	37,650	119,570	3.17
	1.89	35,410	112,100	3.16
	2.97	35,080	109,430	3.12
Average extinction coefficient		35,610	112,160	$\varepsilon_{424}=2050$
		R=0.9985	R=0.9954	$M^{-1}\times cm^{-1}$

With increasing the concentration of mono-substituted 1,2-dihydro-C_{60}-fullerenes (see Table 1.1), the extinction coefficients of mono-substituted 1,2-dihydro-C_{60}-fullerenes at both wavelengths are decreasing. However, their changes are more significant at $\lambda \sim 260$ nm that results in decreasing the relation A_{258}/A_{327}. Such behavior of the solutions can be explained by the formation of associates with variable composition.

1.3.2 SPECTRAL CHARACTERISTICS OF METHANOFULLERENES WITH DIFFERENT NUMBER OF SUBSTITUENTS

We are investigated spectral characteristics of methanofullerenes with the number of substituents in the structure varying from 1 to 5.

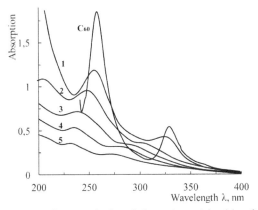

FIGURE 1.2 UV spectra of an unsubstituted C_{60} registered in chloroform and adduct I with different numbers of substituents n (mono-(1), di-(2), tri-(3), tetra-(4), penta-(5) were recorded in acetonitrile at 10^{-5} mol/L concentration.

From the spectra given it may be concluded that as the functionality of the compounds characteristic for C_{60} fullerene increases, the absorption maxima at 258 and 329 nm are shifted to a short-wave region, and their intensity reduces. It is probably connected with breaking the symmetry of the fullerene core and reducing the conjugation length. Starting with a di-substituted methanofullerene sample, the peaks in the 330 nm ranges are degenerated to shoulder. Similar absorption spectra were registered for adduct derivatives II, and III.

The hypsochromic shift of the maximum values of the characteristic fullerene absorption bands is directly proportional to the number of the adduct substituents, and is independent of the type and nature of the solvent (chloroform, acetonitrile) (Fig. 1.3).

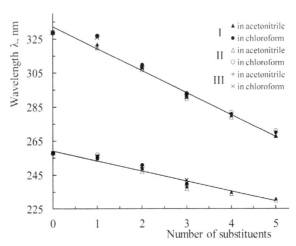

FIGURE 1.3   Wavelength dependence on the number of substituents in I, II and III adducts. Adduct spectra were taken in acetonitrile and chloroform; C_{60} spectra-in chloroform.

As the number of substituents increases, molar absorption coefficients determined in the maxima of the characteristic fullerene absorption bands decrease exponentially (Fig. 1.4). The maximum decrease in the extinction coefficient is observed in transiting from C_{60} to mono- and di-substituted products, slightly changing at $n > 3$.

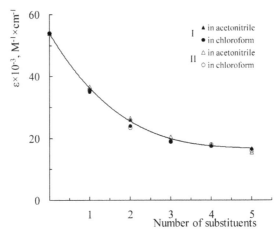

FIGURE 1.4   Dependence of the molar extinction coefficient from the number of substituents in I, II adducts at a wavelength of 329 nm. Adduct spectra were taken in acetonitrile and chloroform, C_{60}-in chloroform.

1.4   CONCLUSION

The UV spectroscopic study revealed that a molar absorption coefficient of a fullerene C_{60} and several synthesized methanofullerenes does not depend on the solvent and functional group nature but hinges on the number of substituents. Moreover, as their number increases, the molar absorption coefficient decreases by the exponential law.

In methanofullerenes spectra a hypsochromic shift of absorption bands maxima occurs as the degree of C_{60} core functionalization increases.

KEYWORDS

- **Extinction Coefficients**
- **Fullerene**
- **Hypsochromic Shift**
- **Methanofullerene**
- **UV-Spectroscopy**

REFERENCES

1. Giacalone, F., & Martin, N. (2006). Chem. Rev, 106, 5136.
2. Makunin, A. V., & Tchecherin, N. G. (2011). Polymer and Nanocarbon Composites from Cosmic Technologies, Ed. M., "Universitetskaya kniga" 150p.
3. Bingel, C. (1993). Chem. Ber., 126, 1957.
4. Camps, X., & Hirsch, A. (1997). J. Chem. Soc. Perkin Trans I., 1595.
5. Grunkin, I. F., & Loginova, N. Yu. (2006). Zh Gen. Chem., 76. 2000.
6. Simone Silvestrini, Daniela Dalle Nogare, & Tommaso Carofiglio, et al. (2011). Eur. J. Org. Chem, 5571.
7. Yves Henri Geerts, Olivier Debever, & Claire Amato, et al. (2009). Beilstein, J. Org. Chem, 5, 49.
8. Torosyan, S. A., Gimalova, F. A., & Mikheev, V. V., et al. (2011). J. Org. Chem, 47(12), 1771.
9. Torosyan, S. A., Gimalova, F. A., & Biglova, Yu. N. (2012). J. Org. Chem, 48(5), 735.
10. Torosyan, S. A., Biglova, Y. N., & Mikheev, V. V. et al. (2012). Mendeleev Communications, 22(4), 199.

CHAPTER 2

BIODEGRADABLE POLYMER FILMS ON LOW DENSITY POLYETHYLENE, MODIFIED CHITOSAN

M. V. BAZUNOVA and R. M. AKHMETKHANOV

CONTENTS

ABSTRACT

Polymer films basing on ultra dispersed low-density polyethylene powder modified by the natural polysaccharide, chitosan have been obtained under the combined effect of high pressure and shear deformation. Their water-absorbent capacity and biodegradability are estimated

2.1 INTRODUCTION

The problem of biodegradability of well-known tonnage industrial polymers is quite urgent for modern studies. It is promising enough to use synthetic and natural polymer mixtures, which can play the roles of both filler and modifier for creating biodegradable environmentally safe polymer materials. The macromolecule fragmentation of the synthetic polymer is to be provided for due to its own biodestruction.

The synthetic polymers have been modified by the natural one under the combined effect of high pressure and shear deformation. The usage of this method for obtaining polymer composites is sure to solve several problems at once. Firstly, the ultra dispersed powders with a high homogeneity degree of the components can be obtained under combined high pressure and shear deformation thus resulting in easing the technological process of production [1]. Secondly, the elastic deformation effects on the polymer material may lead to the chemical modification of the synthetic polymer macromolecules by the natural polymer blocks via recombination of the formed radicals. Thus, it can provide for the polymer product biodegradation. Thirdly, the choice of the best exposure conditions of high pressure and shear deformation on the polymer mixture (modification degrees, process temperature, pressure in the working zone of the dispersant, shear stress values, etc.) may lead to creating environmentally safe biodegradable polymer composite materials processed into products by conventional methods.

Therefore, the working out of the optimal method for obtaining biodegradable polymer films on the basis of ultradispersed powders of LDPE modified by the natural polymer in combined conditions of high pressure and shear deformation is quite expedient. In the paper given a polysaccharide of natural origin, chitosan was used as a polymer.

2.2   EXPERIMENTAL PART

LDPE 10803-020 (90,000 molecular weight, 53% crystallinity degree, and 0,917 g/sm³ density) and chitosan samples of Bioprogress Ltd. (Russia) obtained by alkaline deacetylation of crab chitin (deacetylation degree ~84%), and M_{sd}= 115,000 were used as components for producing biodegradable polymer films.

The initial highly dispersed powders with different mass ratio of components have been obtained by high temperature shearing under simultaneous impact of high pressure and shear deformation in an extrusion type apparatus with a screw diameter of 32 mm [2, 3]. Temperatures in kneading, compression and dispersion chambers amounted to 150°C, 150°C and 70°C, respectively.

The size of particles in powders of LDPE, CTZ and LDPE/CTZ with various mass ratio of the components were determined by "Shimadzu Salid – 7101" particle size analyzer. The film formation was carried out by rotomolding [4] at 135 and 150°C. The film sample thickness amounted to 100 um and 800 um.

The absorption coefficient of the condensed vapors of volatile liquid (water, n-heptane) K' in static conditions is determined by complete saturation of the sorbent by the adsorbent vapors under standard conditions at 20°C [5] and was calculated by the formula: $K' = \frac{m_{absorbedwater}}{m_{sample}} \times 100\%$, where $m_{absorbed\ water}$ is weight of the saturated condensed vapors of volatile liquid, g; m_{sample} is weight of dry sample, g.

Film samples were long kept in the aqueous and enzyme media to determine the water absorption coefficient while the absorbed water weight was calculated. The water absorption coefficient of film samples of LDPE/CTZ with different weight ratio was determined by the formula:

$K = \frac{m_{absorbedwater}}{m_{sample}} \times 100\%$, where $m_{absorbed\ water}$ is water weight absorbed by the sample whereas m_{sample} is the sample weight. Sodium azide was added to the enzyme solution to prevent microbial contamination. Each three days both the water medium and the enzyme solution were changed. The "Liraza" agent of 1 g/dl concentration was used as an enzyme (Immunopreparat SUE, Ufa. Russia).

In experiments for determining the absorption of the condensed vapors of volatile liquid and water absorption coefficients at a confidence level of 0,95 and 5 repeated experiments, the error does not exceed 7%.

The obtained film samples were kept in soil according to the method [6] to estimate the ability to biodegradation. The soil humidity was supported on 50–60% level. The control of the soil humidity was carried out by the hygrometer ETR-310. Acidity of the soil used was close to the neutral with pH = 5.6–6.2 (pH-meter control of 3in1 pH). At a confidence level 0.95 and 5 repeated experiments the experiment error in determining the tensile strength and elongation does not exceed 5%.

Mechanical film properties (tensile strength (σ) and elongation (ε)) were estimated by the tensile testing machine ZWIC Z 005 at 50 mm/min tensile speed.

2.3   RESULTS AND DISCUSSION

It is well known that amorphous-crystalline polymers are subjected to high temperature grinding due to shearing impact on the polymer, for example, a good result is obtained in lowdensity polyethylene at high temperature shearing [2]. Despite chitosan is an infusible polymer, ultra dispersed powder with 6–60 um particles was formed in the output of the rotary disperser after lowdensity polyethylene and chitosan convergence under high pressure and shear deformation (Table 2.1). During high temperature shearing the powders of LDPE and CTZ with the latter not exceeding 60% mass were obtained. The particle distribution of LDPE/CTZ powders does not depend on the ratio of the components of the mixture and little differs from the particle distribution of the powder size of the CTZ under high temperature shearing.

TABLE 2.1   The Absorption Coefficient of the Condensed Water Vapors of Volatile Liquid (water and n-heptane) K> of LDPE/CTZ Powders at 20°C

S. No.	LDPE/CTZ powder, mass. %	Particle size, um	K' by water vapors, %	K' by n-heptane, %
1	0	5.5–8.0; 10.0–80.0	1.10±0.08	17±1
2	20	6.5–63.0	12.3±0.8	11.0±0.8
3	40	6.5–50.0	20±1	5.0±0.4
4	50	4.3–63.0	25±2	4.0±0.3
5	60	6.5–63.0	35±2	4.0±0.3

The speed of the hydrolytic destruction of the polymer materials is closely connected with their ability to water absorption. Values of their absorption capacity according to water and heptane vapors were determined for a number of powder mixture samples of LDPE/CTZ (Table 2.1). It was established that the absorption coefficient of the condensed water vapors is directly proportional to the chitosan content.

As the initial powders, the films with high chitosan content under roto-molding absorb water well (Table 2.2). At the same time thinner films absorb more water for a shorter period of time.

TABLE 2.2 Values of Equilibrium Water Absorption Coefficients K (%) of LDPE/CTZ Films at 20°C.

S . No.	LDPE/ CTZ powder, mass. %	K, %			
		Medium - water		Medium – liraza enzyme (0,1 g/l)	
		Film thickness 100 um	Film thickness 800 um	Film thick-ness 100 um	Film thickness 800 um
1	20	5.0±0.4	2.0±0.2	5.0±0.4	4.0±0.3
2	40	10.0±0.7	4.0±0.3	13.0±0.9	7.0±0.5
3	50	38±3	14±1	40±3	45±3
4	60	–	31±2	–	95.8±0.7

In case the film samples were placed into the enzyme solution, water absorption changes slightly. Firstly, the equilibrium values of the absorption coefficient of films in the enzymatic medium are higher than in water (Table 2.2). It is in the enzymatic medium usage that a longer film exposure (for more than 30–40 days) was accompanied by weight losses of the film samples. Moreover, after 40 days of testing, the film with 50%mass of chitosan and 100um thickness lost its integrity (Fig. 2.1). Films of 800 um thick and chitosan content of 50 and 60% lost their integrity after 2 months of the enzyme agent solution contact. These facts are quite logical as "Liraza" is subjected to a β-glycoside bond break in chitosan. Thus, the destruction of film integrity is caused by the biodestruction process. Higher values of the water absorption coefficient may be explained by enzyme destruction of chitosan chains as well due to some loosening in the film material structure (Table 2.2).

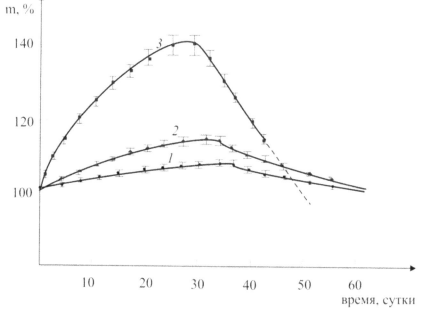

FIGURE 2.1 Curves relative weight change of the film samples LDPE / HTZ (thickness100 micron) HTZ containing 20% (1) 40% (2) and 50% (3) immersed in a solution of the enzyme preparation "Liraz" concentration of 0.11 g/L (at 20°C).

Tests on holding the samples in soil indicate on biodestruction of the obtained film samples either. It is found that the film weight is reduced by 17–18% during the first five months. Here the biggest weight losses are observed in samples with 50–60 mass % of chitosan.

Chitosan introduction into the polyethylene matrix is accompanied by changes in the physical and mechanical properties of the film materials (Table 2.3).

TABLE 2.3   Physical and Mechanical Properties of LDPE/CTZ Film Materials

No.	Chitosan content in LDPE/CTZ, mass. %	σ, MPa		ε, %	
		Film thickness 100 um	Film thickness 800 um	Film thickness 100 um	Film thickness 800 um
1	0	13.30±0.05	40.10±0.05	460.00±0.05	125.00± 0.05

No.	Chitosan content in LDPE/CTZ, mass. %	σ, MPa		ε, %	
		Film thickness 100 um	Film thickness 800 um	Film thickness 100 um	Film thickness 800 um
2	20	5.40±0.05	22.70±0.05	24.30±0.05	13.20±0.05
3	40	7.50±0.05	25.80±0.05	12.50±0.05	7.60±0.05
4	50	11.10±0.05	29.80±0.05	6.00±0.05	6.20±0.05
5	60	11.60±0.05	30.60±0.05	5.20±0.05	4.80±0.05

As seen from Table 2.3, the polysaccharide introduction into the LDPE compounds results in slight decrease in the tensile strength of films. Wherein the number of the chitosan introduced does not affect the composition strength. However, low-density polyethylene/chitosan films obtain much less elongation values as compared with low-density polyethylene films under the same conditions. Thus, films which were obtained on the basis of ultradispersed LDPE powders modified by chitosan possess less plasticity while retain in got her satisfactory strength properties.

2.4 CONCLUSION

A method of obtaining compositions of ultradispersed LDPE powders modified by chitosan under combined high pressure and shear deformation was worked out. The samples received obtain suitable strength properties, a good absorption ability and capability to biodegradation.

KEYWORDS

- **Biodegradable Polymer Films**
- **Chitosan**
- **Low Density Polyethylene**

REFERENCES

1. Bazunova, M. V., Babaev, M. S., Bildanova, R. F., Protchukhan, Yu, A., Kolesov, S. V., & Akhmetkhanov, R. M. (2011). Powder-polymer Technologies in Sorption-Active Composite Materials, Vestn Bashkirs Univer, 16(3), 684–688.
2. Enikolopyan, N. S., Fridman, M. L., Karmilov, A., Yu., Vetsheva, A. S., & Fridman, B. M. (1987). Elastic-Deformation Grinding of Thermo-Plastic Polymers, Reports AS USSR, 296(1), 134–138.
3. Akhmetkhanov, R. M., Minsker, K. S., & Zaikov, G. E. (2006). On the Mechanism of Fine Dispersion of Polymer Products at Elastic Deformation Effects, Plastic masses, 8, 6–9.
4. Sheryshev, M. A. (1989). Formation of Polymer Sheets and Films, Ed. Braginsky, V. A. L., Chemistry Pubishing, 120p.
5. Keltsev, N. V. (1984). Fundamentals of Adsorption Technology, M., Chemistry, 595p.
6. Ermolovitch, O. A., Makarevitch, A. V., Goncharova, E. P., & Vlasova, F. M. 2005. Estimation Methods of Biodegradation of Polymer Materials, Biotechnology, 4, 47–54.

CHAPTER 3

CHEMICAL MODIFICATION OF SYNDIOTACTIC 1,2-POLYBUTADIENE

A. B. GLAZYRIN, M. I. ABDULLIN, O. S. KUKOVINETS, and
A. A. BASYROV

CONTENTS

ABSTRACT

In the paper the interaction of the syndiotactic 1,2-polybutadiene and the reagents of different chemical nature as ozone, peroxy compounds, halogens, carbenes, aromatic amines and maleic anhydride are considered. Various polymer products with a set complex of properties is possible to obtain on the syndiotactic 1,2-polybutadiene basis varying the nature of the modifying agent, a functionalization degree of the polymer and synthesis conditions.

3.1   INTRODUCTION

One of the important directions in the chemistry and technology of macromolecular compounds is the synthesis of new polymeric products through the chemical modification of existing polymers [1].

Physicochemical properties of polymers, their processing conditions, and feasible applications for creating composites of different purposes are in many respects determined by the nature of functional groups contained in polymer macromolecules. The chemical modification of polymers via reactive groups of macromolecules makes it possible to obtain new polymeric products with a wide variety of properties and, thus, to widen the application areas of the modified polymers.

Syndiotactic 1,2-polybutadiene obtained by stereospecific butadiene polymerization in complex catalyst solutions [2–7] provides much interest for chemical modification.

In contrast to 1,4-polybutadiens and 1,2-polybutadiens of the atactic structure, the syndiotactic 1,2-polybutadiene exhibit thermoplastic properties combining elasticity of vulcanized rubber and ability to move to the viscous state at high temperatures and be processed like thermoplastic polymers [8–11].

The presence of unsaturated $>C=C<$ bonds in the syndiotactic 1,2-PB macromolecules creates prerequisites for including this polymer into various chemical reactions resulting in new polymer products. Unlike 1,4-polybutadiens, the chemical modification of syndiotactic 1,2-PB is insufficiently studied, though there are some data available [12–15].

A peculiarity of the syndiotactic 1,2-PB produced nowadays is the presence of statistically distributed cis-and trans units of 1,4-diene polymerization [16, 17] in macromolecules along with the order of 1,2-units at polymerization of butadiene −1.3. Their content amounts to 10–16%. Thus, by its chemical structure, syndiotactic 1,2-polybutadiene can be considered as a copolymer product containing an orderly arrangement of 1,2-units and statistically distributed 1,4 polymerization units of butadiene-1,3:

1

Taking into account syndiotactic 1,2-PB microstructures and the presence of $>C=C<$ various bonds in the polydiene macromolecules, the influence of some factors on the polymer chemical transformations has been of interest. The factors in question are determined both by the double bond nature in the polymer and the nature of the substituent in the macromolecules.

In the paper the interaction of the syndiotactic 1,2-PB and the reagents of different chemical nature as ozone, peroxy compounds, halogens, carbenes, aromatic amines and maleic anhydride are considered.

A syndiotactic 1,2-PB with the molecular weight $M_n = (53–72)×10^3$; $M_w/M_n = 1.8–2.2$; 84–86% of 1,2 butadiene units (the rest being 1,4-polimerization units); syndiotacticity degree 53–86% and crystallinity degree 14–22% was used for modification.

3.2 EPOXIDATION

Influence of the double bond polymer nature in the reaction direction and the polymer modification degree is vividly revealed in the epoxidation reaction of the syndiotactic 1,2-PB, which is carried out under peracids (performic, peracetic, meta-chloroperbenzoic, trifluoroperacetic ones) [21–27], *tret*-butyl hydroperoxide [21–23] and other reagents [28] (Table 3.1).

TABLE 3.1 Influence of the Epoxidizing Agent on the Functionalization Degree α of Syndiotactic 1,2-PB and the Composition of the Modified Polymer*

Epoxidizing agent	α, mol. %	Content in the modified polymer, mol.%			
		Epoxy groups		>C=C<bonds	
		1,2-units	1,4- units	1,2- units	1,4- units
**R¹COOH / H₂O₂	11.0–16.0	-	11.0–16.0	84.0	0–5.0
**R²COOOH	32.1	16.1	16.0	67.9	-
**R³COOOH	34.6	18.6	16.0	65.4	-
Na₂WO₄/ H₂O₂	31.0	15.0	16.0	69.0	-
Na₂MoO₄/ H₂O₂	23.7	7.7	16.0	76.3	-
Mo(CO)₆ / t-BuOOH	18.0	18.0	-	66.0	16.0
NaClO	16.0	-	16.0	84.0	-
NaHCO₃ / H₂O₂	16.0	-	16.0	84.0	-

* the polymer with the content of 1,2- and 1,4-units with 84 and 16%, respectively;
** where R¹ – H-, Me-, Et-, CH₃CH(OH)-; R² – м-ClC₆H₄-; R³ – CF₃-.

Depending on the nature of the epoxidated agent and conditions of the reaction [21–27], polymer products of different composition and functionalization degree can be obtained (Table 3.1).

As established earlier [21–24], at interaction of syndiotactic 1,2-PB and aliphatic peracids obtained in situ under hydroxyperoxide on the corresponding acid, the >C=C< double bonds in 1,4 units of macromolecules are mainly subjected to epoxidation (Scheme 1).

Scheme 1

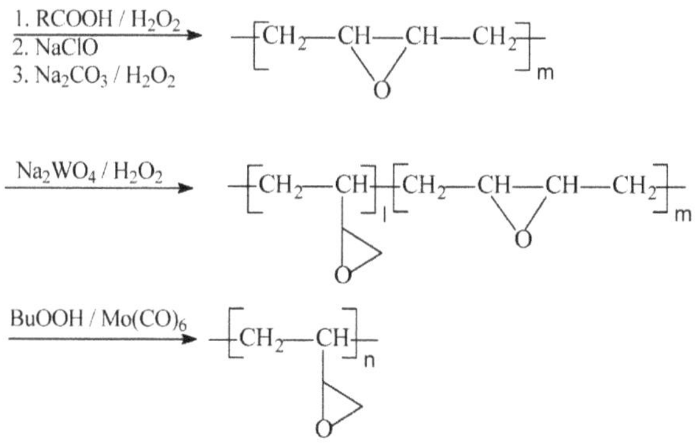

It should be noted that the syndiotactic 1,2-PB epoxidation by the sodium hypochlorite and percarbonic acid salts carried out in the alkaline enables to prevent the disclosure reactions of the epoxy groups and a gelation process of the reaction mass observed at polydiene epoxidation by aliphatic peracids [31, 32]. The functionalization degree of 1,2-PB in the reactions with the stated epoxidizing agents ($\alpha \leq 16\%$, Table 3.1) is determined by the content of inner double bonds in the polymer.

To obtain the syndiotactic 1,2-PB modifiers of a higher degree of functionalization (up to 35%) containing oxirane groups both in the main and side chain of macromolecules (Scheme 1) it is necessary to use only active epoxidizing agents (meta-chloroperbenzoic (MCPBA), trifluoroperacetic acids [23], and metal complexes of molybdenum and tungsten, obtained by reacting the corresponding salts with hydroperoxide) [25] (Table 3.1). From the epoxidizing agents given a trifluoroperacetic acid is most active (Fig. 3.1) [23].

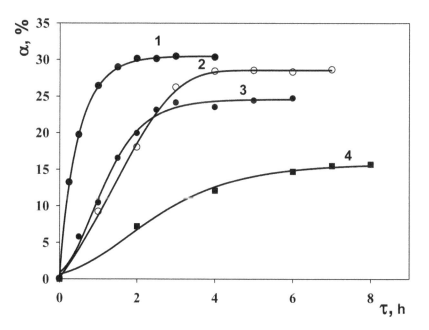

FIGURE 3.1 Influence of the peracid nature on the kinetics of oxirane groups accumulation at the syndiotactic 1,2-PB epoxidation: 1 – TFPA ([Na_2HPO_4] / [TFPA] = 2; 0°C); 2 – MCPBA (50°C); 3 – Na_2MoO_4 / H_2O_2 (55°C); 4 – HCOOH / H_2O_2 (50°C).

However, at the syndiotactic 1,2-PB epoxidation by the trifluoroperacetic acid, a number of special conditions is required to prevent the gelation of the reaction mass, namely usage of the base (Na_2HPO_4, Na_2CO_3 et al.) and low temperature (less than 5°C) [23].

At reacting the syndiotactic 1,2-PB and the catalyst complex [t-BuOOH - $Mo(CO)_6$] a steric control at approaching the reagents to the double polymer bond is carried out [21,22,25]. This results in participation of less active but more available vinyl groups of macromolecules in the reaction (Table 3.1, Scheme 1).

Thus, modified polymer products with different functionalization degrees (up to 35%) may be obtained on the syndiotactic 1,2-PB basis according to the epoxidizing agent nature. The products in question contain oxirane groups in the main chain (with aliphatic peracids, percarbonic acid salts, and NaClO as epoxidizing agents), in the side units of macromolecules [t-BuOOH - $Mo(CO)_6$] or in 1,2- and 1,4-units (trifluoroperacetic acid, MCPBA, metal complexes of molybdenum and tungsten).

3.3   OZONATION

At interaction of syndiotactic 1,2-PB and ozone the influence of the inductive effect of the alkyl substituents at the carbon-carbon double bond on the reactivity of double bonds in 1,2 and 1,4-units of butadiene polymerization is vividly displayed. Ozone first attacks the most electron-saturated inner double bond of the polymer chain. The process is accompanied by the break in the $>C=C<$ bonds of the main chain of macromolecules and a noticeable decrease in the intrinsic viscosity and molecular weight of the polymer (Fig. 3.1) at the initial stage of the reaction (functionalization degree $\alpha<10\%$) [17–20]. Due to the ozone high reactivity, partial splitting of the vinyl groups is accompanied by spending double bonds in the main polydiene chain. However, it does not affect the average molecular weight of the polymer up to the functionalization degree of 15% (Fig. 3.2).

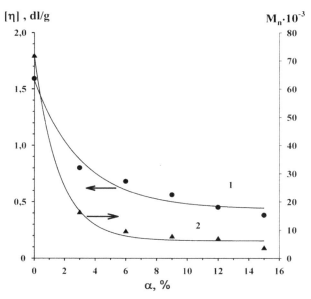

FIGURE 3.2 Dependence of the intrinsic viscosity [η] (1) and the average molecular weight M_n (2) of the formyl derivative of the syndiotactic 1,2-PB from the functionalization degree of the α polymer (with chloroform as a solvent, 25°C).

Depending on the chemical nature of the reagent used for the decomposition of the syndiotactic 1,2-PB ozonolysis products (dimethyl sulfide or lithium aluminum hydride) [17–20], the polymer products containing aldehyde or hydroxyl groups are obtained (Scheme 2).

Scheme 2

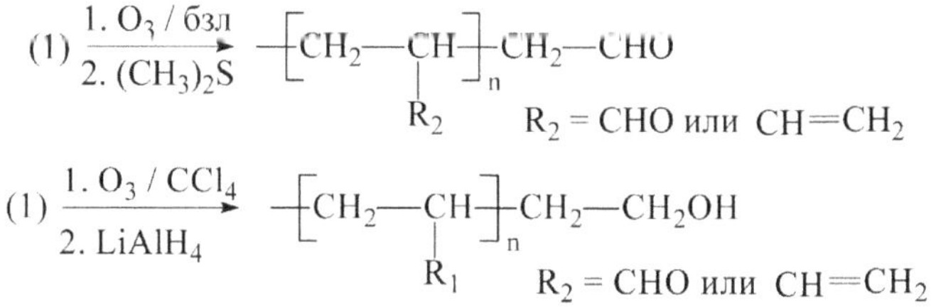

$$(1) \xrightarrow[\text{2. (CH}_3)_2\text{S}]{\text{1. O}_3 \text{ / бзл}} -[CH_2-CH]_n-CH_2-CHO$$

$$R_2 \qquad R_2 = CHO \text{ или } CH=CH_2$$

$$(1) \xrightarrow[\text{2. LiAlH}_4]{\text{1. O}_3 \text{ / CCl}_4} -[CH_2-CH]_n-CH_2-CH_2OH$$

$$R_1 \qquad R_2 = CHO \text{ или } CH=CH_2$$

The structure of the modified polymers is set using IR and NMR spectroscopy [17]. The presence of C-atom characteristic signals connected with aldehyde (201.0–201.5 ppm) or hydroxyl (56.0–65.5 ppm) groups in ^{13}C NMR spectra allows identifying the reaction products.

Thus, the syndiotactic 1,2-PB derivatives with oxygen-contained groupings in the macromolecules may be obtained via ozonation. Modified 1,2-PB with different molecular weight and functionalization degrees containing hydroxi- or carbonyl groups are possible to obtain by regulating the ozonation degree and varying the reagent nature used for decomposing the products of the polymer ozonation.

3.4 HYDROCHLORINATION

In adding hydrogen halides and halogens to the $>C=C<$ double bond of 1,2-PB, the functionalization degree of the polymer is mostly determined by the reactivity of the electrophilic agent. Relatively low degree of polydiene hydrochlorination (10–15%) at interaction of HCl and syndiotactic 1,2-PB [16, 39, 40] is caused by insufficient reactivity of hydrogen chloride in the electrophilic addition reaction by the double bond (Table 3.2). Due to this, more electron-saturated $>C=C<$ bonds in 1,4 units of butadiene polymerization are subjected to modification.

TABLE 3.2 Influence of the Syndiotactic 1,2-PB Hydrochlorination Conditions on the Functionalization Degree α and Chlorine Content in the Modified Polymer (20–25°C; HCl Consumption – 0.2 mol/h per mol 1,2-PB; [AlCl$_3$]=5 mas. %)

Solvent	Reaction time, h	Chlorine content in the polymer, mas. %	α, %	Polymer output, %
chloroform*	24	4.1	10.5	92.0
dichloroethane*	24	5.9	15.1	90.3
chloroform	24	12.6	32.2	91.1
dichloroethane	14	18.8	47.9	92.9
dichloroethane	18	25.9	66.3	95.1
dichloroethane	24	27.9	71.2	96.7

*w/o catalyst.

The process is intensified at polydiene hydrochlorination under AlCl$_3$ due to a harder electrophile H$^+$[AlCl$_4$]$^-$ formation at its interaction with HCl [40]. In this case double bonds of both 1,2- and 1,4-polidiene units take part in the reaction (Scheme 3):

Scheme 3

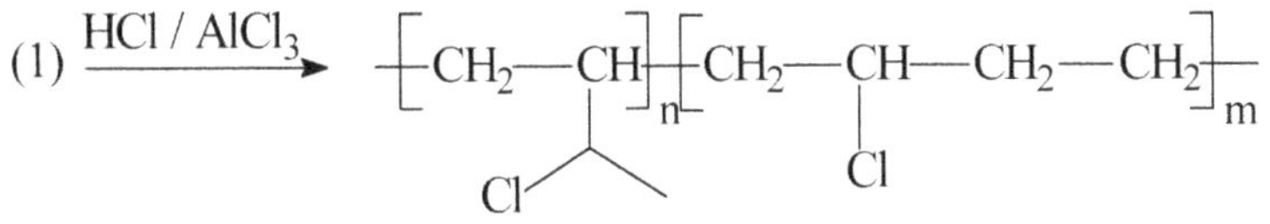
(1)
HCl / AlCl3
CH2
CH
n
CH2
CH
CH2
CH2
m
Cl
Cl

Usage of the catalyst $AlCl_3$ and a polar solvent medium (dichloroethane) allows speeding up the hydrochlorination process (Table 3.2) and obtaining polymer products with chlorine contained up to 28 mass.% and the functionalization degree α up to 71% [40].

By the ^{13}C NMR spectroscopy method [40] it is established that the *>C=C<* double bond in the main chain of macromolecules is more active at 1,2 PB catalytic hydrochlorination. Its interaction with HCl results in the formation of the structure (a) (Scheme 4). At hydrochlorination of double bonds in the side chain the chlorine atom addition is controlled by formation of the most stable carbocation at the intermediate stage. This results in the structure (b) with the chlorine atom at carbon β-atom of the vinyl group (Scheme 4):

Scheme 4

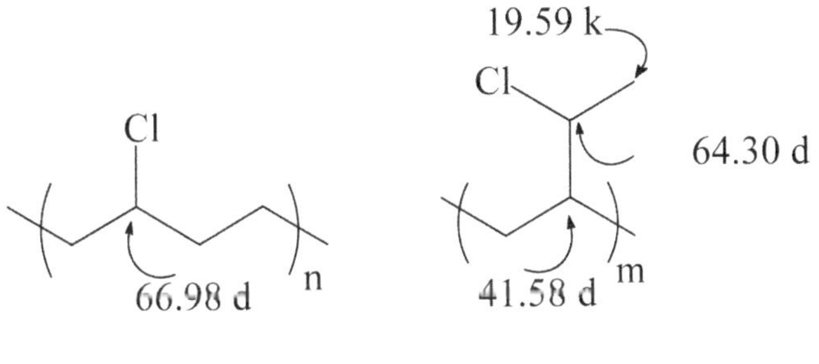
Cl
66.98 d
n
19.59 k
Cl
64.30 d
41.58 d
m

a *b*

3.5 HALOGENATION

Effective electrophilic agents like chlorine and bromine easily join double carbon-carbon bonds [16, 41–43] both in the main chain of syndiotactic 1,2-PB and in the side chains of macromolecules (Scheme 5):

Scheme 5

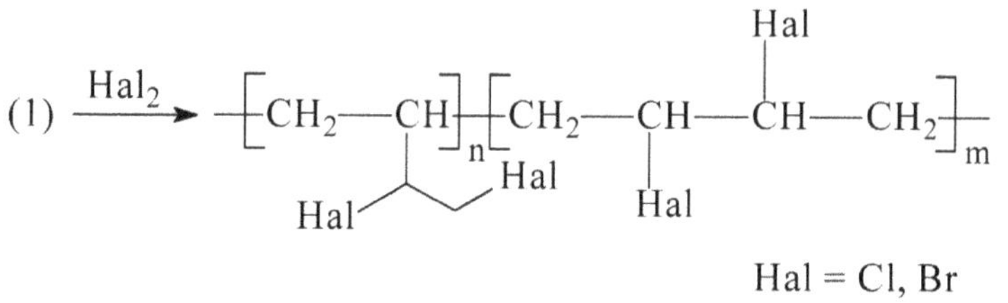

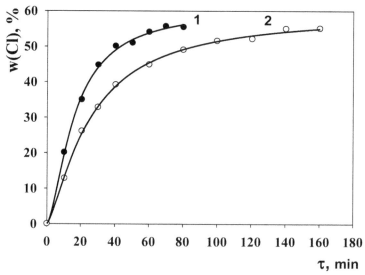

$$(1) \xrightarrow{Hal_2} \left[CH_2-CH \right]_n \left[CH_2-CH-\overset{Hal}{\underset{Hal}{CH}}-CH_2 \right]_m$$

Hal = Cl, Br

The reaction proceeds quantitatively: syndiotactic 1,2-PB chlorine de-rivatives with chlorine w(Cl) up to 56 mass.% (α ~98% (Fig. 3.3) and syndiotactic 1,2-PB bromo derivatives with bromine up to 70 mass. % (α ~ 94%) are obtained.

FIGURE 3.3 Kinetics of syndiotactic 1,2-PB chlorination. Chlorine consumption: *1* - 1 mol/h per mol of syndiotactic 1,2-PB; *2* - 2 mol/h per mol of syndiotactic 1,2-PB; 20°C, with CHCl₃ as a solvent.

According to the ^{13}C NMR spectroscopy, polymer molecules with di-halogen structural units (Scheme 6) and their statistic distribution in the macro chain serve as the main products of syndiotactic 1,2-PB halogena-tion [16, 43].

Scheme 6

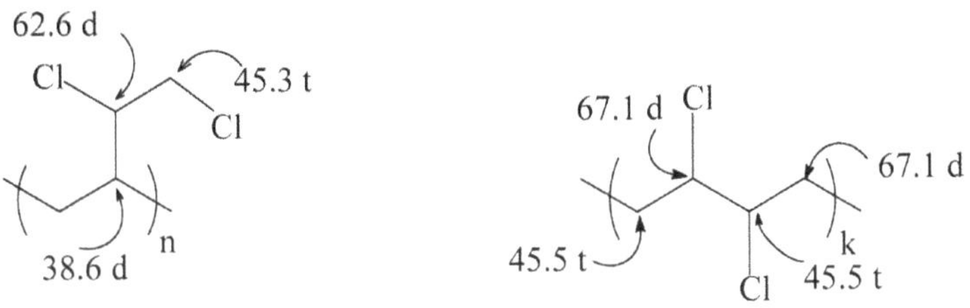

d   e

3.6 DICHLOROCYCLOPROPANATION

There is an alternative method for introducing chlorine atoms into the 1,2-PB macromolecule structure, namely a dichlorocyclopropanation reaction. It is based on generating an active electrophile agent in the reaction mass at dichlorocarbene modification, which is able to interact with double carbon-carbon polymer bonds [45–47]. The syndiotactic 1,2-PB dichlorocyclopropanation is quite effective at dichlorocarbene generating by Macoshi by the chloroform reacting with an aqueous solution of an alkali metal hydroxide. The reaction is carried out at the presence of a phase transfer catalyst and dichlorocarbene addition in situ to the double polydiene links [48–51] according to Scheme 7:

Scheme 7

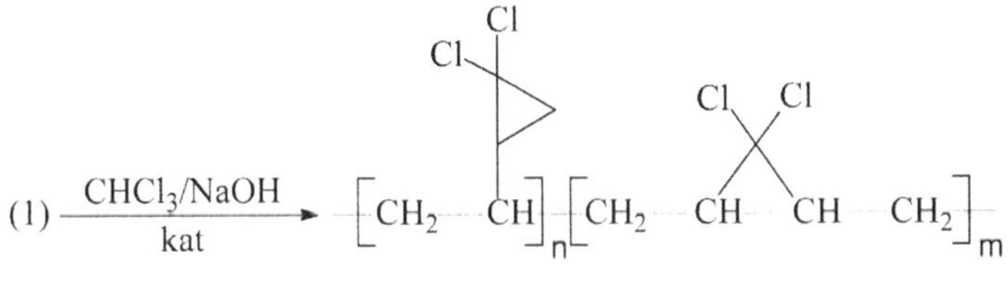

The ^{13}C NMR spectroscopy results testify the double bonds dichlorocyclopropanation both in the main chain and side chains of polydiene macromolecules (Scheme 8). Cis-and trans-double bonds in the 1,4-addition units [48] are more active in the dichlorocarbene reaction.

The polymer products obtained contain chlorine up to 50 mass.% which corresponds to the syndiotactic 1,2-PB functionalization degree ~ 97%, that is, in the reaction by the Makoshi method full dichlorocarbenation of unsaturated $>C=C<$ polydiene bonds is achieved [49,51].

Scheme 8

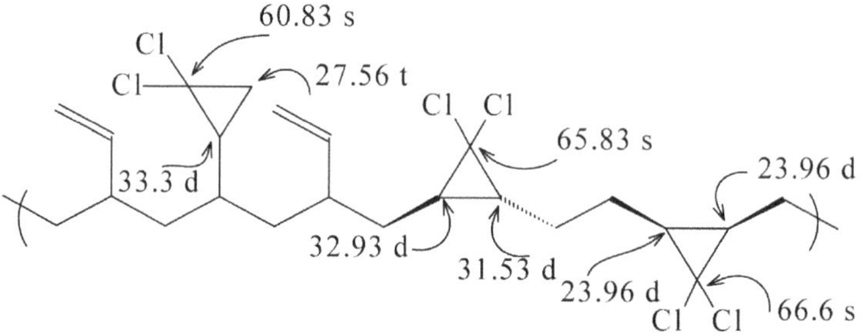

3.6   INTERACTION WITH METHYLDIAZOACETATE

Modified polymers with methoxycarbonyl substituted cyclopropane groups [50–54] are obtained by interaction of syndiotactic 1,2-PB and carb methoxycarbonyl generated at a catalytic methyldiazoacetate decomposition in the organic solvent medium (Scheme 9):

Scheme 9

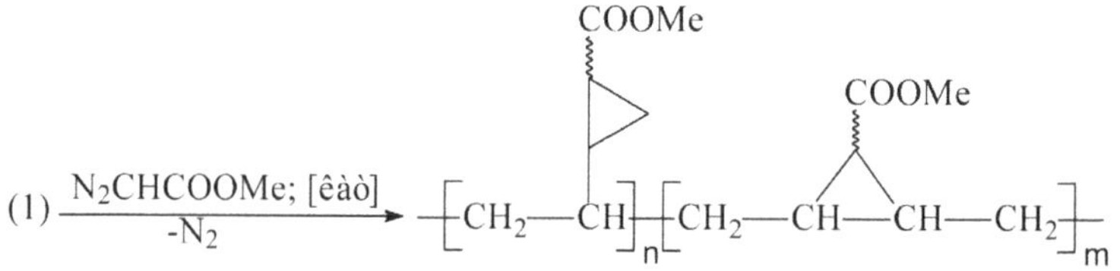

The catalytic decomposition of alkyldiazoacetates comprises the formation of the intermediate complex of alkyldiazoacetate and the catalyst [55, 56]. The generated alcoxicarbonylcarbene at further nitrogen splitting [57] is stabilized by the catalyst with the carbine complex formation, interaction of which with the alkene results in the cyclopropanation products (Scheme 10):

Scheme 10

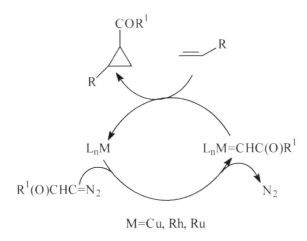

M=Cu, Rh, Ru

The output of the cyclopropanation products is determined by the reactivity of the $>C=C<$ double bond in the alkene as well as the stability and reactivity of the carbine complex $L_nM=CH(O)R^1$ which fully depend on the catalyst used [55, 56].

The catalysts applied in the syndiotactic 1,2-PB cyclopropanation range as follows: $Rh_2(OAc)_4$ (α=38%) > [Cu OTf]× 0.5 C_6H_6 (α=28%) > $Cu(OTf)_2$ (α=22%) [50].

By the ^{13}C NMR spectroscopy methods it is established that in the presence of copper (I) (II) compounds, double bonds both of the main chain and in the side units of syndiotactic 1,2-PB macromolecules are subjected to cyclopropanation whereas at rhodium acetate $Rh_2(OAc)_4$ mostly the $>C=C<$ bonds in the 1,4-addition units undergo it [51].

Thus, catalytic cyclopropanation of syndiotactic 1,2-PB under methyldiazoacetate allows obtaining polymer products with the functionalization degree up to 38% and their macromolecules containing cyclopropane groups with an ester substituent. The determining factor influencing the cyclopropanation direction and the syndiotactic 1,2-PB functionalization degree is the catalyst nature. By using catalysts of different chemical nature it is possible to purposefully obtain the syndiotactic 1,2-PB derivatives containing cyclopropane groups in the main chain (with rhodium acetate as a catalyst) or in 1,2- and 1,4-polydiene units (copper compounds), respectively.

Along with the electronic factors determined by different electron saturation of the $>C=C<$ bonds in 1,2- and 1,4-units of polydiene addition and the catalyst nature used in modification, the steric factors may also influence the reaction and the syndiotactic 1,2-PB functionalization degree. The examples of the steric control may serve the polydiene reactions with aromatic amines and maleic anhydride apart from the above considered epoxidation reactions of syndiotactic 1,2-PB by *tret*-butyl hydroperoxide.

3.7   INTERACTION WITH AROMATIC AMINES

Steric difficulties prevent the interaction of double bonds of the main chain of syndiotactic 1,2-PB macromolecules and aromatic amines (aniline, N,N-dimethylaniline and acetanilide). In the reaction with amines [17, 19, 48] catalyzed by $Na[AlCl_4]$ the vinyl groups of the polymer enter the reaction and form the corresponding syndiotactic 1,2-PB arylamino derivatives (Scheme 11):

Scheme 11

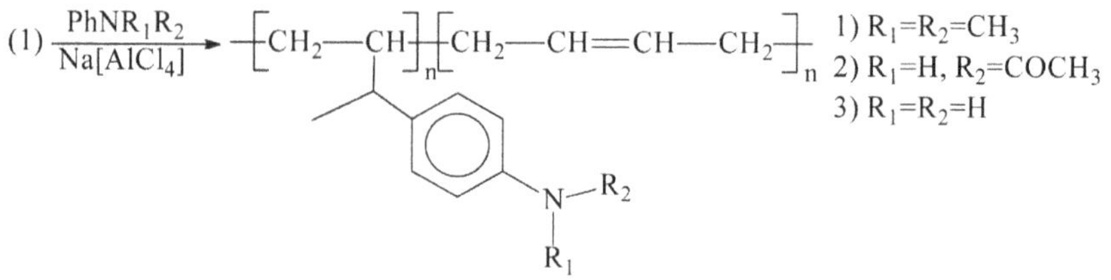

From the NMR spectra analysis it is seen that the polymer functionalization is held through the β-atom of carbon vinyl groups [17].

Introduction of arylamino groups in the syndiotactic 1,2-PB macromolecules leads to increasing the molecular weight M_w (Fig. 3.4) and the size of macromolecular coils characterized by the mean-square radius of gyration $\left(\overline{R^2}\right)^{1/2}$ » [17, 20].

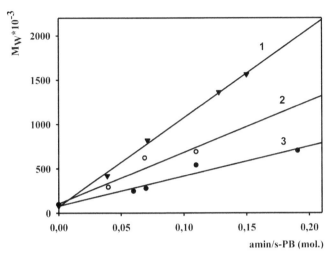

FIGURE 3.4 Influence of the aromatic amine nature on the molecular weight (M_w) of the polymer modified by: 1 – acetanilide; 2 – N,N-dimethylaniline; 3 – aniline.

The results obtained indicate to the intramolecular interaction of monomer units modified by aromatic amines with vinyl groups of polydiene macromolecules at syndiotactic 1,2-PB modification. This leads to the formation of macromolecules of the branched and linear structure (Scheme 12):

Scheme 12

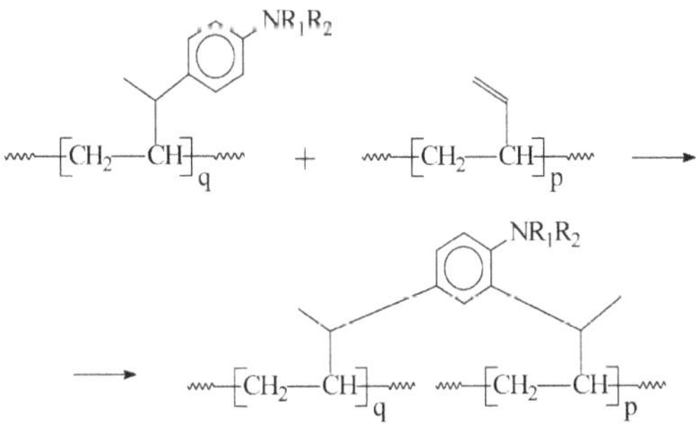

Steric difficulties determined by the introduction of bulky substituents in the polydiene units ("a neighbor effect" [58, 59]) does not allow to obtain polymer products with high functionalization degree as the aryl-amino groups in the modified polymer does not exceed 8 mol.%. At the same time secondary intermolecular reactions are induced in the synthesis process involving arylamino groups of the modified macromolecules and result in the formation of linear or branched polymer products with high molecular weight.

3.8   INTERACTION WITH MALEIC ANHYDRIDE

The polymer products with anhydride groups are synthesized by thermal adding (190°C) of the maleic anhydride to the syndiotactic 1,2-PB [48, 50] (Scheme 13):

Scheme 13

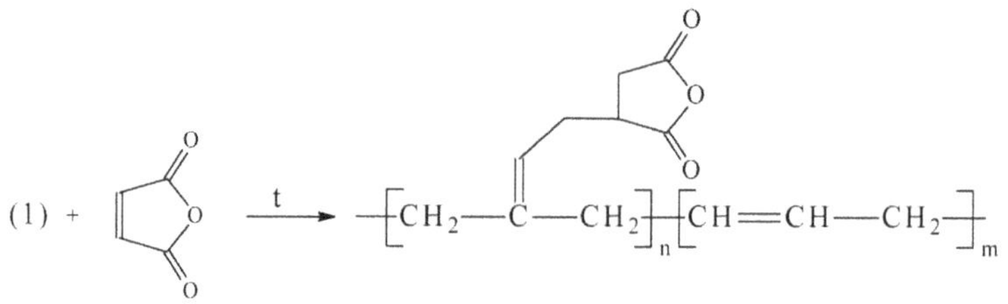

The ^{13}C NMR spectroscopy results show that the maleic anhydride addition is carried out as an ene-reaction [59–61] by the vinyl bonds of the polymer without the cycle disclosure and the double bond is moved to the β-carbon atom of the vinyl bond [55,57]. The maleic anhydride addition to the $>C=C<$ double bonds of 1,4-units of polydiene macromolecules does not take place. As in synthesis of the arylamino derivatives of syndiotactic 1,2-PB, it is connected with steric difficulties preventing the interaction of bulk molecules of the maleic anhydride with inner double bonds of the polymer chain [56].

At syndiotactic 1,2-PB modification by the maleic anhydride, the so-called "neighbor effect" is observed, i.e. the introduction of bulk substituents into the polymer chain prevents the functionalization of the neighboring polymer units due to steric difficulties. For this reason the content of anhydride groups in the modified polymer molecules do not exceed ~15 mol. %.

Thus, $>C=C<$ double bonds in 1,2- and 1,4-units of syndiotactic 1,2-PB macromolecules considerably differ in the reactivity due to the polydiene structure. The inductive effect of the alkyl substituents resulting in the increase of the electron density of the inner double bonds of macromolecules determines their high activity in the considered reactions with different electrophilic agents.

At interaction of syndiotactic 1,2-PB with strong electrophiles (ozone, halogens, dichlorocarbene) both inner double bonds and side vinyl groups of polydiene macromolecules are involved in the reaction. It results in polymer products formation with quite a high functionalization degree. In the case when the used reagent does not display enough activity (interaction of syndiotactic 1,2-PB and hydrogen chloride and aliphatic peracids), the process is controlled by electronic factors: more active double bonds in 1,4-units of the polymer chain are subjected to modification whereas the formed polymer products are characterized by a relatively low functionalization degree.

Polymer modification reactions are mostly carried out through vinyl groups at appearance of steric difficulties. They are connected with formation of a bulky intermediate complex or usage of reagents of big-sized molecules (reactions with aromatic amines, maleic anhydride, t-BuOOH / $Mo(CO)_6$). Such reactions are controlled by steric factors. The predominant course of the reaction by the side vinyl groups of polymer macromolecules is determined by their more accessibility to the reagent attack. However, in such reactions high functionalization degree of syndiotactic 1,2-PB cannot be achieved due to steric difficulties arousing through the introduction of bulky substituents into the polymer chain. They limit the reagent approaching to the reactive polydiene bonds.

Thus, a targeted chemical modification of the polydiene accompanied by obtaining polymer products of different content and novel properties can be carried out using differences in the reactivity of $>C=C<$ double bonds of syndiotactic 1,2-PB. Various polymer products with a set complex of properties is possible to obtain on the syndiotactic 1,2-PB basis varying the nature of the modifying agent, a functionalization degree of the polymer and synthesis conditions.

KEYWORDS

- **Chemical Modification**
- **Functionalization Degree**
- **Polymer Products**
- **Syndiotactic 1,2-Polybutadiene**

REFERENCES

1. Kochnev, A. M., & Galibeev, S. S. (2003). Khimiya i Khimicheskaya Tekhnologiya, 46(4), 3–10.
2. Byrihina, N. N., Aksenov, V. I., & Kuznetsov, E. I. (2001). Patent RF 2177008.
3. Ermakova, I., Drozdov, B. D., Gavrilova, L. V., & Shmeleva, N. V. (1998). Patent, R. F., 2072362.
4. Luo Steven, (2002). Patent US 6284702.
5. Wong Tang Hong, & Cline James Heber (2000). Patent US 5986026.
6. Ni Shaoru, Zhou Zinan, & Tang Xueming (1983). Chinese Journal of Polymer Science, 2, 101–107.
7. Monteil, V., Bastero, A., & Mecking, S. (2005) Macromolecules, 38, 5393–5399.
8. Obata, Y., Tosaki, Ch., & Ikeyama, M. (1975). Polym. J., 7(2), 207–216.
9. Glazyrin, A. B., Sheludchenko, A. V., Zaboristov, V. N., & Abdullin, M. I. (2005). Plasticheskiye Massy, 8, 13–15.
10. Abdullin, M. I., Glazyrin, A. B., Sheludchenko, A. V., Samoilov, A. M., & Zaboristov, V. N. (2007). Zhurnal prikladnoy khimii, 80(11), 1913–1917.
11. Xigao, J. (1990). Polymer Sci. 28(9), 285–288.
12. Kimura, S., Shiraishi, N., & Yanagisawa, S. (1975). Polymer-Plastics Technology and Engineering, 5(1), 83–105.
13. Lawson, G. (2003). Patent US 4960834.
14. Gary, L. (2001). Patent US 5278263.
15. Dontsov, A. A., Lozovik, G. Y. (1979). Chlorinated Polymers, Chemistry, Moscow, 232p.
16. Asfandiyarov, R. N. (2008). Synthesis and Properties of Halogenated 1, 2-Polybutadienes, Ph.D. Chem. Science–Ufa, Bashkir State University, 135p.
17. Kayumova, M. A. (2007). Synthesis and Properties of the Oxygen-and Aryl-containing Derivatives of 1, 2-Syndiotactic Polybutadiene Crystal, Ph.D, Chem. Science-Ufa, Bashkir State University, 115p.
18. Abdullin, M. I., Kukovinets, O. S., Kayumova, M. A., Sigaeva, N. N., Ionov, I. A., Musluhov, R. R., & Zaboristov, V. N. (2004). Vysokomolek.Soyed, 46(10), 1774–1778.
19. Gainullina, T. V., Kayumova, M. A., Kukovinets, O. S., Sigaeva, N. N., Muslukhov, R. R., Zaboristov, V. N. & Abdullin, M. I. (2005). Polymer Sci. Ser.B., 47(9), 248–252.

20. Abdullin, M. I., Kukovinets, O. S., Kayumova, M. A., Sigaeva, N. N., & Musluhov, R. R. (2006). Bashkirskiy khimicheskiy Zhurnal, 13(1), 29–30.
21. Gainullina, T. V., Kayumova, M. A., Kukovinets, O. S., Sigaeva, N. N., Muslukhov, R. R., Zaboristov, V. N., & Abdullin, M. I. (2005). Vysokomolek. Soyed, 47(9), 1739–1744.
22. Abdullin, M. I., Gaynullina, T. V., Kukovinets, O. S., Khalimov, A. R., Sigaeva, N. N., Musluhov, R. R., & Kayumova, M. A. (2006). Zhurnal Prikladnoy Khimii, 79(8), 1320–1325.
23. Abdullin, M. I., Glazyrin, A. B., Kukovinets, O. S., & Basyrov, A. A. (2012). Khimiya i Khimicheskaya Tekhnologiya, 55(5), 71–79.
24. Abdullin, M. I., Glazyrin, A. B., Kukovinets, O. S., Dokichev, V. A., & Basyrov, A. A. (2013). Izvestiya Ufimskogo Nauchnogo Tsentra Rossiyskoy Akademii Nauk, 1, 29–37.
25. Valekzhanin, I. V., Abdullin, M. I., Glazyrin, A. B., Kukovinets, O. S., & Basyrov, A. A. (2012). Actual Problems of Humanitarian and Natural Sciences, 3, 13–14.
26. Kalimullina, G. I., Abdullin, M. I., Glazyrin, A. B., Kukovinets, O. S., & Basyrov, A. A. (2012). Actual Problems of Humanitarian and Natural Sciences, 5, 15–18.
27. Abdullin, M. I., Glazyrin, A. B., Kukovinets, O. S., & Basyrov, A. A. (2012). Mezhdunarodnyy Nauchno-issledovatelskiy Zhurnal, 4, 36–39.
28. Abdullin, M. I., Glazyrin A. B., Kukovinets, O. S., Valekzhanin I. V., Klysova, G. U., & Basyrov, A. A. (2012). Patent RF. 2465285.
29. Abdullin, M. I., Glazyrin, A. B., Kukovinets, O. S., Valekzhanin, I. V., Kalimullina, R. A., & Basyrov, A. A. (2012). Patent RF 2456301.
30. Kurmakova, I. N. (1985). Vysokomolek. Soyed 21(12), 906–910
31. Hayashi, O., Kurihara, H., & Matsumoto, Y. (1985). Patent US 4528340.
32. Blackborow, & John, R. (1991). Patent US 5034471.
33. Jacobi, M. M., Viga, A., & Schuster, R. H. (2002). Raw Materials and Application. 3, 82–89.
34. Emmons, W. D., & Pagano, A. S. J. (1955). Am. Chem. Soc. 77(1), 89–92.
35. Xigao Jian, Allan, S., & Hay, J. (1991). Polym. Sci. Polym. Chem. Ed. 29, 1183–1189.
36. Abdullin, M. I., Glazyrin, A. B., & Asfandiyarov, R. N. (2009). Vysokomolek. Soyed, 51(8), 1567–1572.
37. Glazyrin, A. B., Abdullin, M. I., Muslukhov, R. R., & Kraikin, V. A. (2011). Polymer Science, Ser. A., 53(2), 110–115.
38. Abdullin, M. I., Glazyrin, A. B., Akhmetova, V. R., & Zaboristov, V. N. (2006). Polymer Science Ser. B. 48(4), 104–107.
39. Abdullin, M. I., Glazyrin, A. B., & Asfandiyarov, R. N. (2007). Zhurnal Prikladnoy khimii, 10, 1699–1702.
40. Abdullin, M. I., Glazyrin, A. B., Asfandiyarov, R. N., & Akhmetova, V. R. (2006). Plasticheskiye Massy, 11, 20 22.
41. Lishanskiy, I. S, Shchitokhtsev, V. A., & Vinogradova, N. D. (1966). Vysokomolek, Soyed, 8, 186–171.
42. Komorski, R. A., Horhe, S. E., & Carman, C. J. (1983). J. Polym. Sci Polym. Chem, Ed. 21, 89–96.
43. Nonetzny, A., & Biethan, U. (1978). Angew Makromol Chem. 74, 61–79.

44. Abdullin, M. I., Glazyrin, A. B., Kukovinets, O. S., Valekzhanin, I. V., Klysova, G. U., Basyrov, A. A., Patent R.F. 2462478.
45. Glazyrin, A. B., Abdullin, M. I., & Kukovinets, O. S. (2009). Vestnik Bashkirskogo Universiteta, 14(3), 1133–1140.
46. Glazyrin, A. B., Abdullin, M. I., & Muslukhov, R. R. (2012). Polymer Science, Ser B, 54, 234–239.
47. Abdullin, M. I., Glazyrin, A. B., Kukovinets, O. S., Basyrov, A. A., & Muslukhov, R. R. (2012). Khimiya i Khimicheskaya Tekhnologiya, 55(5), 71–78.
48. Glazyrin, A. B., Abdullin, M. I., Khabirova, D. F., & Muslukhov, R. R. (2012). Patent RF 2456303.
49. Glazyrin, A. B., Abdullin, M. I., Sultanova, R. M., Dokichev, V. A., & Muslukhov, R. R. (2012). Patent RF 2443674.
50. Glazyrin, A. B., Abdullin, M. I., Sultanova, R. M., Dokichev, V. A., Muslukhov, R. R., Yangirov, T. A., & Khabirova, D. F. (2012). Patent RF, 2445306.
51. Glazyrin, A. B., Abdullin, M. I., Sultanova, R. M., Dokichev, V. A., Muslukhov, R. R., Yangirov, T. A., & Khabirova, D. F. 2012. Patent RF 2447055.
52. Shapiro Ye, A., Dyatkin, A. B., & Nefedov, O. M. (1992). Diazoether. Nauka, Moscow, 78p.
53. Gareyev, V. F., Yangirov, T. A., Kraykin, V. A., Kuznetsov, S. I., Sultanova, R. M., Biglova, R. Z., & Dokichev, V. A. (2009). Vestnik Bashkirskogo Universiteta, 14(1), 36–39.
54. Gareyev, V. F., Yangirov, T. A., Volodina, V. P., Sultanova, R. M., Biglova, R. Z., & Dokichev, V. A. (2009). Zhurnal Prikladnoy Khimii, 83(7), 1209–1212.
55. Fedtke, M. (1990). Chemical Reactions of Polymers, Chemistry Moscow, 152p.
56. Kuleznev, V. N., & Shershnev, V. A. (2007). The Chemistry and Physics of Polymers, Kolos, Moscow, 367p.
57. Vatsuro, K. V., & Mishchenko, G. L. (1976). Named Reactions in Organic Chemistry, Khimiya, Moscow 528p.
58. Lanov, K. O. (1985). Maleic Acid and Maleic Anhydride, Khimiya, Moscow, 163p.
59. Popovich, T. D., Glovati, O. L., Pliyev, T. N., & Gordash, Y. T. (1976). Neftekhimiya, 16(5), 778–784.

CHAPTER 4

APPROACH TO CREATING PROLONGED DRUGS ON CARBON POLYMER BASIS

M. V. BAZUNOVA and S. V. KOLESOV

CONTENTS

ABSTRACT

Methods of obtaining multilayer particles of nanocarbon as an inner core, an ultrathin shell from zerovalent silver covered by a soluble polyvinyl-pyrrolidone layer have been worked out. They can be used for creating the carrier systems aimed at targeted delivery of prolonged drugs.

4.1 INTRODUCTION

Drugs are to be distributed in the body so as to get to the place of action only and application of nanostrucutural carriers for the targeted drug delivery seems to be a way out. As the unique nanoparticle ability consists in its extremely developed surface, the nanosystems of drug delivery help to overcome low solubility and unsatisfactory sorption properties of the latest medicines [1, 2]. Thus, the problem of working out biocompatible nanosystems for targeted drug delivery seems to be quite topical.

Nanosystems are a kind of spheric nanoparticles of a multilayer structure with a monolythic inner core and ultra-thin shell from different medicines covered by the polymer layer. It is necessary to use quite an affordable nanocarbon with its high sorption capacity to form the core [3]. This may provide for consolidation and retention of drugs and protective polymer layers. The usage of metal shells of highly dispersed silver particles is quite promising for tumor diagnostics and targeted delivery of anticancer medicines. Biocompatible water-soluble polymers as polyvinylpyrrolidone may serve as polymer shells of a prolonged action

Therefore, the obtaining of multilayer particles with a nanocarbon as an inner core and an ultra-thin shell of zerovalent silver or a drug covered by the soluble polymer is viewed as quite expedient.

4.2 EXPERIMENTAL PART

Statistic sorption capacity of nanocarbon as related to silver ions was determined by the Ag^+ concentration changes in the solution before and after keeping with the nanocarbon bunching attachment of a definite mass at stirring during 30–120 min at 20°C. The Ag^+ concentration control in the solution was carried out by the Folgard method [4].

The method of chemical reduction in adsorption layers with sodium borohydrideas a reducing agent was used to obtain the ultra-thin shell of the zerovalent silver on the nanocaron surface [5].

The nanocarbon statistic sorption capacity as related to the N-vinyl-pyrrolidone and polyvinylpyrrolidone was determined by their concentration changes in the solution before and after keeping with the nanocarbon bunching attachment of a definite mass at stirring during 30–120 min at 20°C.

The N-vinylpyrrolidone and polyvinylpyrrolidone control in aqueous solutions was carried out by SHIMADZU- UV 2450PC. This method allows determining vinylpyrrolidone content in mixtures with ±1.5rel. % precision.

Polymerization of N-vinylpyrrolidone was carried out by the known method [6] in a hydroalcoholic solution under hydrogen peroxide adding to ammonia.

The experimental error does not exceed 5% in all weight methods at 0.95 confidence level and three repeated experiments.

4.3   RESULTS AND DISCUSSION

In the paper given the affordable nanocarbon obtained by the oxidative methane condensation with 200 m^2/g specific surface and 50–60 nm average particle diameter is used as one of the main components of the worked out functional materials [7].

The silver cation sorption by nanocarbon from the silver nitrate was investigated. It is established that the nanocarbon sorption capacity by Ag^+ amounts to 0.69 g/g which 30 times exceeds the activated carbon sorption capacity by Ag^+ [4].

Poly N-vinylpyrrolidone usage is most effective as a polymer shell of the worked out multilayer nanosystems. Polyvinylpyrrolidone is a polymer soluble in water and other polar solvents. It is widely used as a binding substance in production of tableted drugs as it prolongs other agent action and forms complexes with other substances [8].

Thus, the nanocarbon sorption activity was studied as related to N-vinylpyrrolidone. The spectrophotometric method was used to determine the content of vinylpyrrolidone in the solution (Fig. 4.1, Table 4.1). The content changes of vinylpyrrolidone in the solution after nanocarbon sorption at its different initial concentrations (3%, 5% and 10%) were studied.

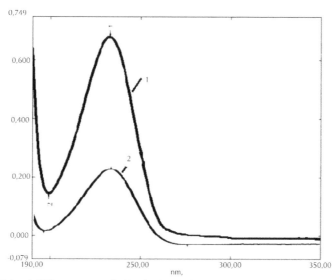

FIGURE 4.1 1-UV spectrum of 5% vinylpyrrolidone solution before sorption (20°C); 2- UV spectrum of 5% vinylpyrrolidone solution after sorption (20°C).

TABLE 4.1 Nanocarbon Sorption Capacity to Vinylpyrrolidone after Sortion for 30 Min at Different Initial Vinylpyrrolidone Concentrations, T = 20°C

Initial concentration of vinylpyrrolidone, %	3	5	10
Sorption capacity of nanocarbon to vinylpyrrolidone after sorption during 30 min, g/g	0.32	0.70	0.56

The data received led to the conclusion that the biggest change in the vinylpyrrolidone concentration after nanocarbon sorption is observed at 5% initial vinylpyrrolidone concentration.

The obtaining of the polyvinylpyrrolidone water soluble shell on the surface of the nanocarbon modified by silver nanoparticles serves the example for non-covalent modifications a result of π–π interaction through the polymer wrapping of nanoparticles (Fig. 4.2).

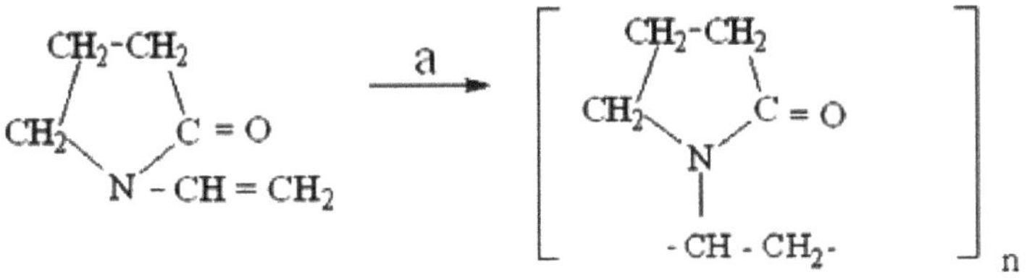

FIGURE 4.2   N-vinylpyrrolidone polymerization scheme: solvent H2O : C2H5OH (20 : 1) , t = 20°C, the initiating system : H2O2 (2.5×10–3 mol/L) : NH3 (5×10–3 mol/L), Ar current.

There are various experimental ways for obtaining multilayer nanoparticles on the nanocarbon basis modified by silver and a water-soluble polyvinylpyrrolidoneshell. The following scheme seems to be the most expedient for possible regulating the particle sizes of the reduced silver and polyvinylpyrrolidone coating on the received nanoparticles:

Ag^+ ions sorption by nanocarbon $\rightarrow$ Ag^+ chemical reduction $\rightarrow$ vinylpyrrolidone sorption $\rightarrow$ vinylpyrrolidone polymerization.

The spectrophotometric method was used for determining the residual content of vinylpyrrolidone after polymerization in the aqueous phase. It was found that the vinylpyrrolidone content in the solution was reduced to 10%.

The UV-spectrum data show the polymer presence (polyvinylpyrrolidone) in the dry sample of the modified nanocarbon (Table 4.2).

TABLE 4.2   Characteristic Frequencies of Absorption of Certain Atom Groups in the UV-Spectrum of Nanocarbon Modified by the Silver Shell and Polyvinylpyrrolidone

Group	v, sm^{-1}
-C-N-	1360–1000
-C-N<	1450–1400
>C=O	1900–1580

The intense band of stretching vibrations of C=O-group of the pyrrolidone ring of 1680 cm^{-1} is displayed in the UV spectrum and is absent at 1630 sm^{-1} for the stretching vibrations of C=C-bonds. It confirms the break of double bonds in N-vinylpyrrolidone while proving the polymer presence.

A sample experiment of Ag^+ reducing in the polyvinylpyrrolidone solution as a stabilizer was carried out to determine a possible size of the reduced silver particles. It is found that the silver particle diameter in freshly prepared sols does not exceed 10–15 nm.

4.4 CONCLUSION

Thus, the worked out methods for obtaining multilayer particles with a nanocarbon as an inner core, an ultrathin shell of the zerovalent silver covered by the soluble polyvinylpyrrolidone can be used for creating the carrier systems for targeted delivery of prolonged drugs.

KEYWORDS

- **Nanocarbon**
- **Polyvinylpyrrolidone**
- **Sorption**

REFERENCES

1. Pomogaylo, A. D., Rozenberg, A. S., & Uflyand, I. E. (2000). Metal Nanoparticles in Polymers, M, Chemistry, 672p.
2. Kononova, E. A. (2010). Obtaining Criostability, Adsorption and Bacterium Properties of Ag, Au, Ag, Au Nanoparticles of Ashes and the Carriers. Thesis Abstract. Cand. Chem. Sciences, M., 24p.
3. Bazunova, M. V., Babaev, M. S., Vildanova, R. F., Protchukhan, Yu, A., Kolesov, S. V., & Akhmetkhanov, R. M. (2011). Powder-polymer Technologies in Sorption Active Composite Materials Vestn. Bashkirs. Univer., 16(3), 684–688.
4. Pyatnitskiy, I. V., & Sukhan, V. V. (1975). Analytical Chemistry of Silver. Nauka, M., 289p.
5. Kuzmina, L. N. Obtaining of Silver Particles by Chemical Reduction, J. Russ. Chem. Society, 8, 7–12.
6. Sidelkovskaya, F. P. (1970). Chemistry of N-vinylpyrrolidone and its Polymers, Nauka, M., 150p.
7. Industrial Patent RF 2287543 from 20.11.2006.
8. Pomogaylo, A. D. (1997). Stabilization of Colloid Disperses by Polymers, Success in Chemistry, 66(8), 750–791p.

CHAPTER 5

USAGE OF SPEED SEDIMENTATION METHOD FOR ENZYME DEGRADATION OF CHITOSAN

E. I. KULISH, V. V. CHERNOVA, S. V. KOLESOV, and
I. F. TUKTAROVA

CONTENTS

ABSTRACT

A possibility of using the viscosimetry and sedimentation and diffusion methods in studying the process of enzyme chitosan hydrolysis is discussed in the article. It is shown that a change in the intrinsic viscosity of chitosan may be determined by both the hydrolysis process in the glycoside bonds and transformation of the supramolecule structure of the polymer. Thus, for setting the hydrolysis process it is necessary to use an absolute method for molecular weight determination.

5.1 INTRODUCTION

Chitosan, a polysaccharide of natural origin, as any other natural polymer is capable of biodegradation under the influence of enzyme agents, that is, degradation of the main chain by glycosidic linkages (accompanied by decreasing of its molecular weight). The degradation process of chitosan may be carried out not only under specific enzymes (chitinases and chitosanases), but under the influence of some nonspecific enzymes like lysozyme [1] and celloviridin [2]. If the Mark-Kuhn-Houwink equation is quite known as applied to the studied polymer, then the most convenient method for degradation determination is a viscosimetry method [3–5]. The reduction of polymer viscosity in the solution implicitly indicates to decreasing the chitosan molecular weight during the process of degradation. However, the values of molecular weight received by the viscosimetry method do not match the real ones as this method is relative and requires the usage of gauge dependences [6]. Moreover, the solution viscosity is determined by the macromolecular coil size whereas the coil size and its molecular weight are not one and the same. Thus, reducing a polymer solution viscosity may be testified not just by a degradation process but a transformation of the polymer supramolecular structure in the solution. This transformation may be caused by disintegration of chitosan macromolecule associates [7]. Therefore, it is impossible to definitely name the reason for viscosity reduction basing on the viscosimetric method only. For revealing the reasons for viscosity changes of polymer solutions there is proposed an absolute method for determining the molecular weight, namely a method combining sedimentation and diffusion [8].

5.2 EXPERIMENTAL PART

Three chitosan samples, CHT-1 (Chimmed Ltd., Russia), CHT-2 and CHT-3 (Bioprogress CJSC.) obtained by alkaline deacetylation of crab chitin and enzyme agents "Collagenase," "Liraza" ("Immunopreparat" SUE, Ufa) and "Tripsin" ("Microgen" FSUE SPA, Omsk) were chosen as objects for investigation. The chitosan 2% (mass.) solution was prepared by 24 h dissolution at ambient temperature. Acetic acid of 1 g/dl concentration was used as a solvent. 5% chitosan mass of the enzyme agent predissolved in a small amount of water was introduced in the polymer solution. Enzymatic destruction was carried out for 20 days in 25 °C. Sodium azide (0.04% from the chitosan mass) was added to the chitosan solution to prevent microbial contamination [9].

The intrinsic viscosity of chitosan was determined in a buffer solution of 0.3 m. of acetic acid and 0.2 m of sodium acetate, whereas the relative viscosity was established for polymer solutions of 1% acetic acid by the standard method [8] in 25 °C.

After exposition with the enzyme, the chitosan was precipitated, washed by distilled water and dried to a constant weight. The sedimentation S_c and diffusion D_c coefficients were calculated from the sedimentation data received by the analytical ultracentrifuge MOM 3180 (Hungary) in 546 nm wave length and $25 + 0.1°C$ temperature with Philpot-Svensson optics. The angle of the phase plate (Philpot-Svensson's inclination angle) in both cases amounted to 30°. The rotor speed varied depending on the task set. The diffusion coefficients D_c for 4–5 concentrations in the range of 0.15–0.40 g/dL were found by the erosion rate of the solvent-solution boundary in time. The experiments were carried out in the two-sector cell in a rotor speed of 6000 rpm. The diffusion coefficients D_c were calculated in relation to the square Q under the gradient curve to the maximum ordinate H in the moment of time t:

$$D_c = (Q/H)^2 /4\pi t.$$

The sedimentation coefficient S_c within the same concentration range was in time determined by the movement rate of the solvent-solution boundary. The experiments were conducted in a two-sector cell in a rotor speed of 45,000 rpm. The S_c value was calculated by the formula:

$$S_c = (dx/dt)/\omega^2 t,$$

where x – a maximum coordinate of the curve of the gradient in the sedimentation boundary segment (sm),

t – time (sec),

$\omega = 2\pi n/60$ – angular velocity of rotation of the rotor,

t – rotor speed per minute.

M_{SD} value was calculated by the found S_0 and D using the first Swedberg formula [8]:

$M_{SD} = (S_0 / D) [(RT/(1-v\rho_0)]$.

where R – gas constant; T – absolute temperature, K; v – partial specific gravity of the polymer in a solution, sm^3/g; ρ_0 – density of the solvent g /sm^3. S_0 and D are sedimentation and diffusion constants received by S_c and D_c extrapolation on the zero concentration.

All concentration S_c and D_c dependences were linear which allowed carrying out the S_c and D_c value extrapolation on the zero concentration, finding true values of these characteristics (S_0 and D) and calculating the true value of M_{SD} accordingly [8].

X-ray phase investigations were fulfilled on the X-ray diffractometer "Shimadzu XRD 6000." CuK-radiation wavelength 0.154 nm, initial angle 5.00 deg., measuring step 0.05 deg., the final angle 40.00 deg., was used.

5.3   RESULTS AND DISCUSSION

In the earlier researches [10, 11] a considerate decrease in relative viscosity of chitosan solutions in acetic acids in the presence of enzyme agents "Liraza," "Collagenase" and "Tripsin" was established (Fig. 5.1, curves 1–3). Taking into account that acids are chitosan solvents, the reason for viscosity reduction in chitosan solutions may consist in the process of acid degradation rather than the influence of a nonspecific enzyme. Some decrease in relative viscosity of chitosan really occurs even in the absence of enzyme agents (Fig. 5.1, curve 4).

Moreover, exposure of chitosan solutions with enzyme agents is accompanied by the decrease in characteristic viscosity values of its solutions (Fig. 5.2, curves 1–3). While no enzymes are used, the exposure of chitosan in the acetic acid also leads to drastic reduction in single values of its characteristic viscosity (Fig. 5.2, curve 4).

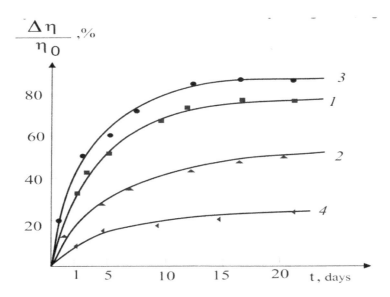

FIGURE 5.1 The dependence of relative viscosity changes of 2% chitosan solution from the exposure time with the enzyme agents "Liraza" (1), "Collagenase" (2) and "Tripsin" (3) and in the absence of enzyme agents (4); 1% acetic acid concentration. The chitosan:enzyme weight ratio (%) is 95:5.

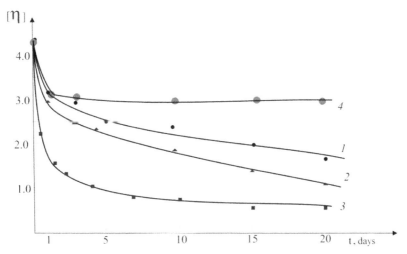

FIGURE 5.2 The dependence of characteristic viscosity changes from the time of chitosan exposure in the solution of CHT-1 separated from the 1% acetic acid solution in the absence of enzyme agents (4) and with enzyme agents "Tripsin" (1), "Liraza" (2), and "Collagenase" (3).

So the characteristic viscosity of CHT-1 failing to pass through the dissolving stage in the acetic acid was determined in an acetate buffer (pH=4.5, [η] = 4.45 dl/g). If this chitosan sample is sustained in a 1% acetic acid solution for 24 h and separated from the solution, its characteristic viscosity is determined in an acetate buffer and the received value [η] amounts to 3.1 dl/g which is considerably less than the initial value. If the polymer is sustained in the acetic acid solution for a longer period of time, it does not lead to further reduction of characteristic viscosity. Similar regularities are observed in all studied samples of chitosan. As seen from the table, after chitosan is precipitated from the acetic acid solution, its characteristic viscosity always decreases. Additional heating of the sample is also accompanied by some reduction in viscosity values. The received experimental data may be described from the positions of the acid degradation process. However, the application of the method combining high-speed sedimentation and diffusion for investigating chitosan solutions in the acetic acid allows to register the following results (see the Table 5.1). In the absence of enzyme agents the values of the sedimentation constant and the diffusion coefficient coincide with the values of the source chitosan samples both for the precipitated and heated ones. This definitely indicates that the value of the chitosan molecular weight during the process of its reprecipitation and heating does not change. Thus, dissolution of chitosan in the acetic acid solution is not accompanied by polymer degradation.

TABLE 5.1   Some Physical and Chemical Characteristics of Chitosan Samples in an Acetic Buffer (pH=4.5)

Chitosan sample	Used enzyme agent	[η], dl/g	$S_0 \times 10^{13}$, s	$D \times 10^7$, sm²/s	$M_{SD} \times 10^{-5}$
CHT-1	–	4.45	3.00	1.30	1.60
CHT-1[2]	–	3.69	3.10	1.35	1.57
CHT-1[1]	–	3.50	3.15	1.38	1.59
CHT-1[3]	"Liraza"	1.10	1.84	4.00	0.25
CHT-1[3]	"Collagenase"	0.60	1.79	5.70	0.21
CHT-1[3]	"Tripsin"	1.80	1.97	3.50	0.42
CHT-2	–	6.10	7.13	1.02	3.34
CHT-2[1]	–	4.90	7.20	1.10	3.12

TABLE 5.1 *(Continued)*

Chitosan sample	Used enzyme agent	$[\eta]$, dl/g	$S_0 \times 10^{13}$, s	$D \times 10^7$, sm²/s	$M_{SD} \times 10^{-5}$
CHT-3	–	3.45	2.15	1.49	0.99
CHT-3[1]	–	2.82	2.20	1.56	0.97

Note:

[1] – a chitosan sample precipitated from the solution in the acetic acid;

[2] – a chitosan sample precipitated from the solution in the acetic acid and subjected to additional heating by 70–80°C in the solution of the acetic buffer;

[3] – a chitosan sample separated from the enzyme-containing solution of 1% acetic acid, holding time of chitosan in the solution is 20 days and the enzyme content is 5% from the chitosan mass.

On the contrary, chitosan holding in an acetic acid solution in the presence of the enzymes leads to considerable reduction in the sedimentation constant and increase of the diffusion coefficient (see the Table 5.1). Thus, it results in significant reduction of the chitosan molecular weight taking place during the biodegradation process.

The observed changes in the values of characteristic and intrinsic viscosity in the absence of enzyme agents are probably connected with structural transformations such as destruction of chitosan associates in the acetic acid solution. As a XRD analysis shows, the source chitosan and the one dissolved in the acetic acid have different supramolecular organization (Fig. 5.3).

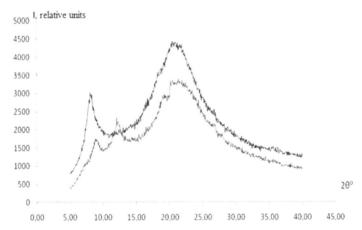

FIGURE 5.3 X-ray pattern of the source CHT-3 (1) and a film sample received from the chitosan solution in the 1% acetic acid (2).

5.4 CONCLUSION

The usage of the method of speed sedimentation and diffusion allows to state that any changes of chitosan molecular weight do not occur in the absence of enzyme agents. So the observed reduction in intrinsic viscosity of chitosan after the dissolving in the acetic acid is determined by transformation of the initial supramolecular structure of the polymer. On the contrary, the decrease of the molecular weight of the polymer takes place under the influence of the degradation enzymes of the chitosan main chain in the presence of the enzyme agents.

KEYWORDS

- **Chitosan**
- **Enzyme Degradation**
- **Sedimentation**

REFERENCES

1. Skryabina, K. G., Vikhoreva, G. A., & Varlamova, V. P., Eds., Chitin and Chitosan, Production, Properties and Application, Nauka, (2002), 365.
2. Ilyina, A. V., Tkatcheva, Yu, V., & Varlamov, V. P. (2002). Applied Biochemistry and Microbiology, 38(2), 132.
3. Mullagaliev, I. R., Artuganov, G. E., & Melentiev, A. I. (2006). In Modern Perspectives in Chitin and Chitosan Researches, 305.
4. Vikhoreva, G. A., Rogovina, S. Z., Pchelko, O. M., & Galbrakh L. S. (2001). Polym Sci., B ser., 43(6), 1079.
5. Fedoseeva, E. N., Semtchikov, Yu, D., & Smirnova, L. A. (2006). Polym. Sci, B., Ser, 48(10), 1930.
6. Budtov, V. P. (1992). Physical Chemistry of Polymer Solutions, Petersburg, S., 384.
7. Kulish, E. I., Chernova, V. V., Vildanova, R. F., Bolodina, V. P., & Kolesov, S. V. (2011). Bulletin of Bashkir State University, 16(3), 681.
8. Rafikov, S. R., Budtov, V. P., & Monakov, Yu, B. (1978). Introduction to Physics and Chemistry of Polymer Solutions, Moscow, 320.
9. Martirosova, E. I., Plaschina, I. G., Feoktistova, N. A., Ozhimkova, E. V., & Sidorov, A. I. (2009). All-Russia Scientific Conference of Topical Problems of Modern Science and Education. URL: http://e-conf.nkras.ru/konferencii/Martirosova.pdf.

10. Chernova, V. V., Kulish, E. I., Volodina, V. P., & Kolesov, S. V. (2008). Modern Perspectives in Chitin and Chitosan Researches, 9th International Conference. Moscow, 234.
11. Kulish, E. I., Chernova, V. V., Volodina, V. P., Torlopov, M. A., & Kolesov, S. V. (2010). Modern Perspectives in Chitin and Chitosan Researches, 10th International Conference, Moscow, 274.

CHAPTER 6

STUDYING OF ANTIBIOTICS RELEASE FROM MEDICINAL CHITOSAN FILMS

E. I. KULISH and A. S. SHURSHINA

CONTENTS

ABSTRACT

Medicinal film coatings on the basis of chitosan and antibiotics both of cephalosporin series and of aminoglycoside one have been considered. It has been shown that the antibiotics release from a film will be determined by the amount of antibiotics connected with chitosan by hydrogen bonds, on the one hand, and by the state of the polymer matrix, on the other.

6.1   INTRODUCTION

The main task of antibacterial chemical therapy is the selective suppression of microorganisms without damaging the organism as a whole. The decrease of the antibiotics therapy efficiency observed is mainly caused by the possibility of origination of bacteria strain tolerant (resistant) to these antibiotics. Polymer derivatives of antibiotics can help to solve this task. We've made an attempt to use chitosan as a carrier of antibacterial preparations. In this situation the choice of chitosan (ChT) as a polymer carrier of a medicinal preparation is not accidental because this polymer possesses a whole spectrum of unique properties making it indispensable in polymer medicine [1]. In the present study we've considered some approaches to creating antibacterial ChT-based coatings of prolonged action suitable for treating surgical, burning and slowly healing wounds of different etiology.

6.2   EXPERIMENTAL PART

The objects of investigation were a ChT specimen produced by the company "Bioprogress" (Russia) and obtained by acetic deacetylation of crab chitin and antibiotics both of cephalosporin series – cephazolin sodium salt (CPhZ), cephotoxim sodium salt (CPhT), and of aminoglycoside series – amikacin sulfate (AMS), gentamicin sulfate (GMS). The investigation of the interaction of medicinal preparations with ChT was carried out according to the techniques described in Refs. [2, 3].

ChT films were obtained by means of casting of the polymer solution in acetic acid onto the glass surface with the formation of chitosan acetate (ChTA). The polymer mass concentration in the initial solution was 2 g/dl. The acetic acid concentration in the solution was 1, 10 and 70 g/dl. Aqueous antibiotic solution was added to the ChT solution immediately before films formation. The content of the medicinal preparation in the films was 0.1 mol/mol ChT. The film thickness in all the experiments was maintained constant and equal to 0.1 mm. The kinetics of antibiotics release from ChT film specimens into aqueous medium was studied spectrophotometrically at the wavelength corresponding to the maximum absorption of the medicinal preparation.

In order to regulate the ChT ability to be dissolved in water the anion nature was varied during obtaining ChT salt forms. So, a ChT-CPhZ film is completely soluble in water. The addition of aqueous sodium sulfate solution in the amount of 0.2 mol/mol ChT to the ChT-CPhZ solution makes it possible to obtain an insoluble ChT-CPhZ-Na_2SO_4 film. On the contrary, a ChT-AMS film being formed at the components ratio used in the process of work, isn't soluble in water. Obtaining a water-soluble film is possible if amikacin sulfate is transformed into amikacin chloride (AMCh). In this case the obtained ChT-AMCh film will be completely soluble in water. Thus, the following film specimens have been analyzed in the investigation: ChT-CPhZ and ChT-CPhT (soluble forms); ChT-CPhZ-Na_2SO_4 (insoluble-in-water form); ChT-AMCh (soluble form); ChT-AMS and ChT-GMS (insoluble-in-water forms).

With the aim of determining the amount of medicinal preparation held by the polymer matrix there was carried out the synthesis of adducts of the ChT-antibiotic interaction in the mole ratio 1:1 in acetic acid solution. The synthesized adducts were isolated by double reprecipitation of the reaction solution in NaOH solution with the following washing of precipitated complex residue with isopropyl alcohol. Then the residue was dried in vacuum up to constant mass. The amount of preparation strongly held by chitosan matrix was determined according to the data of the element analysis on the analyzer EUKOEA – 3000.

6.3   THE RESULTS DISCUSSION

On the basis of the chemical structure of the studied medicinal compounds [4] one can suggest that they are able to combine with ChT forming polymer adducts of two types – ChT-antibiotics complexes and polymer salts produced due to exchange interaction. As a result some quantity of medicinal substance will be held in the polymer chain. The interaction taking place between the studied medicinal compounds and ChT was demonstrated by UV- and IR-spectroscopy data. The interaction energies evaluated by the shift in UV-spectra are about 7–12 kJ/mole, which allows us to speak about the formation of complex ChT-antibiotic compounds by means of hydrogen bonds.

Table 6.1 gives the data on the amount of antibiotics determined in polymer adducts obtained from acetic acid solution.

TABLE 6.1   The Amount of Antibiotics Determined in Reaction Adducts.

CH$_3$COOH, g/dl in the initial solution	The antibiotics used	The amount of antibiotics in reaction adduct, % mass.
1	CPhZ	10.1
	CPhT	15.9
	AMS	61.5
	GMS	59.4
10	CPhZ	5.88
	CPhT	57.5
	AMS	55.8
	GMS	31.3
70	CPhZ	3.03
	CPhT	3.7
	AMS	41.3
	GMS	40.1

Attention should be paid to the fact that the amount of medicinal preparation in the adduct of the ChT-medicinal preparation reaction is considerably higher in the case of antibiotics of aminoglycoside series than in the case of antibiotics of cephalosporin series. This can be connected with the fact that CPhZ and CPhT anions interact with ChT polycation forming salts readily soluble in water. In the case of using AM and GM sulfates because of two-base character of sulfuric acid one may anticipate the formation of water-insoluble "double" salts – ChT-AM or ChT-GM sulfates due to which additional quantity of antibiotics is held on the polymer chain.

Table 6.2 gives the data on the value of the rate of AM and GM release from film specimens formed from acetic acid solutions of different concentrations. The rate was evaluated only for water-insoluble films because at using soluble films the antibiotic release was determined not by medicinal preparation diffusion from swollen matrix but by film dissolving.

TABLE 6.2 Transport Properties of Chitosan Films in Relation to Medicinal Preparation Release.

Acetic acid concentration g/dl	The antibiotics used	Release, % mass./h for chitosan specimens
1	AMS	0.5
	GMS	0.4
10	AMS	0.8
	GMS	0.5
70	AMS	1.5
	GMS	1.3

Attention must be given to the fact of interaction between the rate of antibiotics release from chitosan films and their amount, which is strongly held in ChT chain. For example, at increasing the concentration of acetic acid used as a solvent the amount of medicinal preparation connected with the polymer chain decreases in all the cases considered by us. Correspondingly, the rate of antibiotics release from films insoluble in water, increases.

The influence of the amount of medicinal preparation strongly held in ChT matrix, on the rate of medicinal substance release from the film must be most pronounced at comparing the rates of release of antibiotics of aminoglycoside series and cephalosporin one. However, ChT-CPhZ and ChT-CPhT films are soluble in water while ChT-AMS and ChT-GMS ones do not dissolve in water and it isn't correct to compare them. At ChT transition into insoluble form (by adding sodium sulfate) the rate of release of antibiotics of cephalosporin series decreases considerably (Fig. 6.1, curve 1) as compared with a soluble form but still it is higher than that in the case of antibiotics of aminoglycoside series (Fig. 6.1, curve 2). It should be also noted that the rate of antibiotics release from soluble ChT-CPhZ film (Fig. 6.1, curve 3) is also higher than in the case of ChT-AMCh film (Fig. 6.1, curve 4). Thus, considerable difference between the rate of release of aminoglycoside series antibiotics and that of cephalosporin series antibiotics is evidently explained by the difference in the amount of ChT-antibiotics adduct.

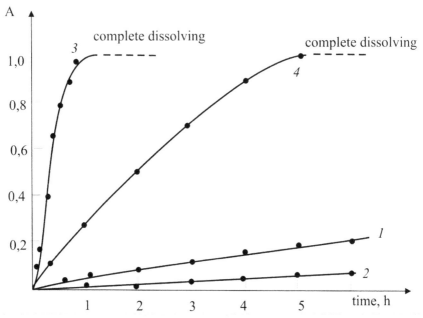

FIGURE 6.1   The kinetic curve of the release of CPhZ (1,3) and AM (2,4) from insoluble (1,2) and soluble (3,4) films.

Thus, at forming film coatings one should proceed from the fact that a medicinal preparation can be distributed in the polymer matrix in two ways. One part of it connected with polymer chain, for example, by complex formation is rather strongly held in that polymer chain. The rest of it is concentrated in polymer free volume (in polymer pores). The rate of release of antibiotics from the film will be determined by the amount of antibiotics connected with ChT by hydrogen bonds, on the one hand, and by the state of the polymer matrix including its ability to dissolve in water, on the other hand.

The work has been carried out due to the financial support of the RFFR and the Republic of Bashkortostan (grant_r_povolzhye_a No. 11-03-97016)

KEYWORDS

- **Chitosan**
- **Medicinal Preparation**
- **Modification**
- **The State of Polymer Matrix**

REFERENCES

1. Skryabin, K. G., Vikhoreva, G. A., Varlamov, V. P. (2002). Chitin and Chitosan, Obtaining Properties and Application, Nauka, M., 365p.
2. Mudarisova, R., Kh, Kulish, E. I., Kolesov, S. V., & Monakov, Yu, B. (2009). Investigation of Chitosan Interaction with Cephazolin, JACh, 82(5), 347–349.
3. Mudarisova, R., Kh, Kulish, E. I., Ershova, N. R., Kolesov, S. V., & Monakov, Yu, B. (2010). The Study of Complex Formation of Chitosan with Antibiotics Amicacin and Hentamicin, JACh, 83(6), 1006–1008.
4. Mashkovsky, M. D. (1997). Medicinal Preparations, Kharkov, Torsing, 2, 278p.

CHAPTER 7

FEATURES OF ENZYMATIC DESTRUCTION OF CHITOSAN IN ACETIC ACID SOLUTION

E. I. KULISH, I. F. TUKTAROVA, and V. V. CHERNOVA

CONTENTS

ABSTRACT

A polymer of natural origin chitosan attracts attention of researchers because it can be used to produce bioactive protecting coatings for treatment surgical wounds and burns. In this case, the biodegradability (enzymatic destruction) of the chitosan material will be affected under the action of nonspecific enzymes of the human body (e.g., hyaluronidase which is present at the wound surface, etc.). The speed of this process will determine the life of the polymer material on the wound surface. Thus, the study of the process of enzymatic destruction of chitosan under the action of nonspecific enzyme preparations will be important, both from scientific and practical points of view. In this paper we determine the kinetic parameters of the process of enzymatic destruction of chitosan by the enzyme hyaluronidase.

7.1   EXPERIMENTAL

The object of investigation was a CHT specimen produced by the company «Bioprogress» (Russia) and obtained by acetic deacetylation of crab chitin with a molecular weight of M_{sh}=113000. As the enzyme preparation was used hyaluronidase enzyme preparation ("Liraze") production of "Microgen" (Moscow, Russia). The concentration of the enzyme preparation was 0.1, 0.2 and 0.3 g/L. Acetic acid of 1 g/ concentration was used as a solvent. CHT concentration in solution ranged from 0.1 to 5 g/dL.

About the process of enzymatic destruction was judged by the falling of the intrinsic viscosity [η] of CHT. The intrinsic viscosity in a solution of acetic acid was determined at 25°C, using the method of Irzhak and Baranov [1]. To determine the initial values of the intrinsic viscosity of CHT solution $[η]_0$ was used at a concentration of c = 0.15 g/dL. To determine the values of intrinsic viscosity during the enzymatic digestion [η], CHT dissolved in acetic acid to which was added a solution of the enzyme preparation maintained for a certain time. Then, a process of enzymatic destruction was quenched by boiling the original solution for 30 min in a water bath. Next, from the initial concentration of the solution c_{ed}, a solution to determine the inherent viscosity at a concentration of c = 0.15 g/dL was made. The process of enzymatic destruction was performed at 36°C.

The initial rate of enzymatic destruction of chitosan V_0 evaluated on the linear part by the drop in its intrinsic viscosity $[\eta]$ and calculated by the formula [2]:

$$V_0 = c_{ed} K^{1/a} ([\eta]^{-1/a} - [\eta]_0^{-1/a})/t \qquad (1),$$

where c_{ed} – the concentration of chitosan in the solution was subjected to enzymatic destruction, g/dL; t – time of destruction, min; K and α – constants of the equation Mark-Houwink $[\eta]=KM^\alpha$; M – molecular weight of chitosan.

To determine the constants in Mark-Houwink equation it is necessary to calculate values of the initial velocity of the enzymatic destruction of the Eq. (1), CHT sample was fractionated into 10 fractions of molecular weight ranging from 20,000 to 150,000 Daltons. The absolute value of the molecular weight fractions of CHT were determined by a combination methods sedimentation velocity and viscosimetry.

The molecular weight of the fractions was determined by the formula:

$$M_{s\eta} = (S_0 \eta_0 [\eta]^{1/3} N_A /(A_{hi}(1-v\rho_0)))^{3/2} \qquad (2),$$

where S_0 – sedimentation constant; η_0 – dynamic viscosity of the solvent, equal to 1.2269×10^{-2} PP; $[\eta]$ – intrinsic viscosity dl/g; N_A – Avogadro number, equal to 6.023×10^{23} mol^{-1}; $(1-v\rho_0)$ – Archimedes factor or buoyancy factor, v – the partial specific volume, cm^3/g, ρ_0 – density of the solvent g/cm^3; A_{hi} – hydrodynamic invariant, equal to 2.71×106.

7.2 RESULTS AND DISCUSSION

Figure 7.1 shows the dependence of the intrinsic viscosity of the CHT solution on the time of standing with the enzyme preparation.

It can be seen, that when increasing time of exposure CHT to the enzyme solution, the viscosity decreases regularly, indicating a reduction of molecular weight CHT. The most significant drop in viscosity occurs in the initial period. Further enzymatic solution CHT exposure, affects the degree of viscosity fall significantly, to a lesser extent that is the reaction rate decreases with time. This course of the kinetic curve is typical for most enzymatic reactions [3]. Increasing the concentration of the enzyme preparation leads to a natural increase in the rate of fall of the intrinsic viscosity.

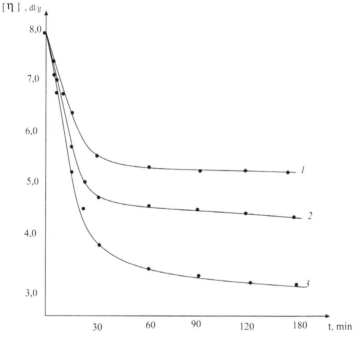

FIGURE 7.1   Dependence of the intrinsic viscosity on the time of standing 1% CHT solution with the enzyme solution with concentration 0.1 (1) 0.2 (2) and 0.3 (3) g/L.

As established in the course of the work, for all the studied solutions for small CHT times degradation decrease the intrinsic viscosity curves are linear. It is on this plot was determined value of the initial rate of enzymatic degradation of V_0, serving as a measure of enzymatic activity of the enzyme with respect to CHT.

To determine the rate of enzymatic degradation of the formula (1) it is needed to determine the value of the constants in the equation of Mark-Houwink. For the sample we used their significance was defined as $\alpha=1{,}02$ and $R=5.57*10^{-5}$.

As studies have shown, the observed dependence of the initial rate of enzymatic destruction of the substrate concentration can be described within the Michaelis–Menten scheme. Figure 7.2 shows the dependence of the initial rate of CHT concentration in solution.

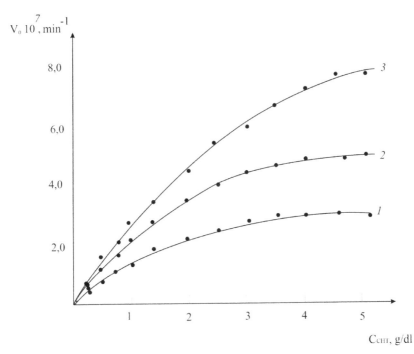

FIGURE 7.2 Dependence of the initial rate of enzymatic destruction of CHT at a concentration of enzyme preparation 0.1 (1) 0.2 (2) and 0.3 (3) g.

It is clear that it has a classic look and occurs a part of a rectangular hyperbola. Submission to double-check the coordinates (a graphical method of Lineweaver-Burk) can accurately determine the value of the Michaelis constant K_m (Table 7.1). As can be seen from the data, its value is constant, independent of the concentration of the enzyme and, in fact characterizes the affinity of the enzyme to the substrate value $V_{max} = k_2 C_e$, where is k_2 – the decay constant of the enzyme-substrate complex, gives a description of the catalytic activity of the enzyme, that is, defines the maximum possible formation of the reaction product at a given concentration of enzyme in an excess of substrate. When the reaction substrate in excess of the maximum rate of reaction depends linearly on the concentration of the enzyme (Fig. 7.3).

TABLE 7.1 The Values of the Constants in the Equation of Michaelis–Menten for CHT Solution in 1% Acetic Acid

C_e, g/l	K_m, g/dL	V_{max}, 10^6
0.1	3.37	0.50
0.2	3.47	0.92
0.3	3.42	1.51

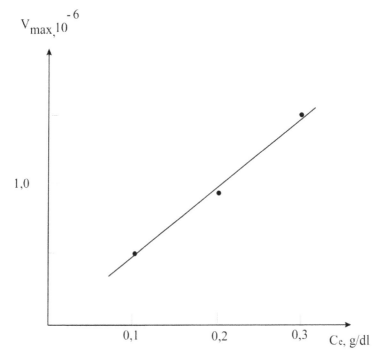

FIGURE 7.3 The dependence of the maximum reaction rate of enzymatic destruction of CHT depending on the concentration of the enzyme preparation in the solution.

7.2 CONCLUSION

Thus, the first time to determine the kinetic characteristics of the enzyme activity of hyaluronidase for the enzymatic destruction of a CHT solution of 1% acetic acid. Certain value of the Michaelis constant K_m » 3.4 g/dL significantly higher than the K_m=0.03 g/dL, as defined in Ref. [1] for the carboxymethylcellulose under the cellulose system of Geotrilium candidum. This fact is obviously due to the fact that the enzyme is used in the nonspecific with respect to CHT and the fact that the conditions of the enzymatic destruction did not meet the temperature and pH optimum of the enzyme hyaluronidase

KEYWORDS

- **Chitosan**
- **Enzymatic Destruction**
- **Hyaluronidase**
- **Viscosity**

REFERENCES

1. Baranov, V., Brestkin, Y., Agranova, S. A., & Pinkevich, V. N. (1986). The Behavior of Macromolecules of Polystyrene in the "Thickened" Good Solvent. Polymer, 28(10), S841–843.
2. Rabinovich, M. L, Klesov, A. A, & Berezin, I. V. (1977). The Kinetics of Enzyme Action Tsellyuliticheskih Geotrilium Candidum, Viscometric Analysis of the Kinetics of Hydrolysis of CMC. Bioorganic Chemistry 3(3), S405–414.
3. Severin, E. S., & Geotar, M., ed., (2004). Biochemistry, Med, 779p.

PART II

NEW METHODS FOR THE SYNTHESIS OF
ISOPRENE—MONOMER FOR THE PRODUCTION
OF ISOPRENE RUBBER

CHAPTER 8

QUANTUM-CHEMISTRY STUDY OF THE FORMATION MECHANISM OF HYDROGENATED FURANS IN TRIFLUOROACETIC ACID BY PRINS REACTION

I. V. VAKULIN, R. R. SYRLYBAEVA, R. F. TALIPOV,
G. R. TALIPOVA, R. R. FAIZULLINA, and A. V. ALLAGULOVA

CONTENTS

ABSTRACT

It is theoretically proved that the mechanism of formation of 3-alkylsub-stituted hydrogenated furans in trifluoroacetic acid by interaction between terminal alkenes and formaldehyde medium which includes recyclization of 3-(2-hydroxyethyl)-1-trifluoromethyl-2,5-dioxolenium ions is preferable.

8.1   INTRODUCTION

β-substituted hydrogenated furans and their derivatives are objects of high practical interest. It shown that these derivatives have anesthetic, anti-septic and antibacterial properties and they exhibit fungicidal and growth regulator activities [1]

The Prins reaction is one of the way to create one-step processes for obtaining derivatives of β-substituted hydrogenated furans with good yields [2–18].

It was shown that 3-alkyl-substituted and 2,3,5-trialkylsubstituted-2,5-dihydrofurans, 4-alkyl-substituted tetrahydrofuranols-3 can be obtained by Prins reaction in trifluoroacetic acid [2–8].

However, the details of the process have not been fully explored. For example, only terminal alkenes enter this reaction. Also the role of the reaction medium is not completely known.

The formation mechanism of hydrogenated furans by coupling between terminal alkenes and aldehydes in trifluoroacetic acid medium described in the literature involves unsaturated alkoxycarbenium ions as the inter-mediate [4–7]. This scheme is proposed by analogy with the mechanism for formation of hydrogenated pyrans under Prins reaction conditions.

$$R^1CH{=}CH_2 \underset{CF_3COOH}{\overset{R^2CHO,\,H^+}{\rightleftarrows}} R^1\overset{+}{C}H{-}CH_2{-}CHR^2{-}OH \xrightarrow{-H} R^1CH{=}CH{-}CHR^2{-}OH \overset{R^2CHO,\,H^+}{\underset{H_2O}{\longrightarrow}}$$

$$\longrightarrow R^1CH{=}CH{-}CHR^2{-}O{-}\overset{+}{C}HR^2 \xrightarrow{lim}$$

$R^1, R^2 = Alk$

Earlier, the steps of producing initial carbocations, which are generated by protonated formaldehyde molecules attacking the double bond, followed by its further transformation into allylic alcohols was investigated using quantum chemical methods. This scheme fits with kinetics data: it has been experimentally determined that the disappearance of every reagent was described by the equation for first order reaction. However, the suggested scheme does not explain some experimental facts: it does not consider non-terminal alkenes or allylic alcohols in the reaction, and the 2,3-dihydrofurans and hemiformals typical for interaction between allylic alcohols with formaldehyde in acid mediums are absent among the reaction products.

Based on data from the literature we may offer an alternative mechanism that proceeds via the formation of dioxolenium ions followed by recyclization by nucleophilic substitution mechanism to produce hydrogenated furans in acid medium [19–23].

In this work, both potentially possible mechanisms of hydrogenated furan formation were investigated, the main attention is focused on the study of the rate-controlling steps of the two schemes. Thermodynamic parameters for these steps, geometries of intermediates and transition states were determined by using quantum chemical methods.

The rate-controlling steps are considered the processes of intramolecular cyclization of unsaturated alkoxycarbenium ions and recyclization of substituted dioxolenium ions with lead to produce tetrahydrofuranium cations. As a model, the following transformations have been examined:

$R_1HC=CR_2CHR_3OCHR_4 \overset{+}{\longrightarrow}$

R_1- R_5, = Alk

1-8   9-16   17-23

R_1, R_2, R_3, R_4 = H (1, 9, 17); R_1 = CH_3, R_2, R_3, R_4 = H (2, 10, 18); R_1 = C_2H_5, R_2, R_3, R_4 = H (3, 11, 19); R_1 = n-C_3H_7, R_2, R_3, R_4 = H (4, 12, 20); R_1 = i-C_3H_7, R_2, R_3, R_4 = H (5, 13, 21); R_1, R_2= CH_3, R_3, R_4 = H (6, 14, 22); R_2 = CH_3, R_1, R_3, R_4 = H (7, 15); R_1 = C_2H_5, R_3, R_4= CH_3, R_2, = H (8, 16, 23).

$R_1 - R_4 = Alk$

24-30   31-37   38-44

R$_1$, R$_2$, R$_3$, R$_4$ = H (24, 31, 38); R$_1$ = CH$_3$, R$_2$, R$_3$, R$_4$= H (25, 32, 39); R$_1$ = C$_2$H$_5$, R$_2$, R$_3$, R$_4$ = H (26, 33, 40); R$_1$, R$_2$ = CH$_3$, R$_3$, R$_4$ = H (27, 34, 41); R$_2$ = CH$_3$, R$_1$, R$_3$, R$_4$ = H (28, 35, 42); R$_3$ = CH$_3$, R$_1$, R$_2$, R$_4$ = H (29, 36, 43); R$_4$ = CH$_3$, R$_1$, R$_2$, R$_3$ = H (30, 37, 44).

8.2   THE CHOICE OF CALCULATION METHOD

Research of chemical reaction mechanisms by methods of quantum chemistry requires accurate definition of structure and energetics of intermediates and transition states which participate in transformations. So total energy calculations for particles was made using compound methods (CM [24–30]) reproducing results for high level MP4/6-311+G(fd,p) approach. The molecular geometries, zero-point energy and entropy were determined by the MP2/6-31G(d,p) approach.

The general form of the suggested CM method can be presented by this way:

E [MP4 $^{MP2\Delta(+fdp)}_{6-311}$]= E$_0$[MP4/6-311G]+ Δ(+,fd,p), -311+G(fd,p)// MP2/6-31G(d,p) approach; E$_0$[MP4/6-311G] – a base level energy; Δ(+,fd,p)=E[MP2/6-311+G(fd,p)] – E[MP2/6-311G] – a correction for basis set extension, K – an empirical correction introducing into the formula in order to increase accuracy of the CM method.

A value of an increment K can be equal to zero (K$_0$) or can be calculated using the following formula: K$_B$ = a·N$_{core}$+b·N$_{val}$+c·N$_{pair}$, where N$_{core}$ is the number of core electrons, N$_{val}$ is the number of valence electrons in the molecule, N$_{pair}$ is the number of unshared electron pairs. Values of coefficients a, b, c and y are estimated with the method of least squares by consideration of the error of the dependence compound method received for G2/97 test set [25] with K$_0$ on the number of corresponding electrons. The method error was defined as the difference between the total energies obtained from MP4/6-311G + (fd, p) and CM calculations of molecules.

We think that the MP4 $^{MP2\Delta(+fdp)}_{6\text{-}311}$ method is optimal for this research. Both a good approximation to results calculated directly at MP4(full)/6-311G+(fd,p) (Table 8.1) and a considerable economy of computer resources (Table 8.2) are provided by its use.

TABLE 8.1   Average Deviations and Maximal Deviations for the MP4 $^{MP2\Delta(+fdp)}_{6\text{-}311}$ Scheme Against Results of MP4/6-311G+(fd,p) Method (kJ/mol).

Types of compounds	Compounds	K_B	
		Average deviation	Maximal deviation
allyloxymethyl cations	2–3.6	1.1	1.5
transition states	10–11, 14	3.2	3.9
tetrahydrofuranium cations	18–19, 22	0.4	0.6

Table 8.2. Relative CPU Time for Some Compounds.

Method	24	31
MP4/6-311G+(fd,p)	43.3	43.2
MP4 $^{MP2\Delta(+fdp)}_{6\text{-}311}$	1	1

Considering the sensitivity to solvents for the studied processes, the effects of the solvent were also investigated for some model reactions. The solvation effects on these reactions were examined with the polarizable continuum model (PCM). PCM calculations were carried out using the MP4 $^{MP2\Delta(+fdp)}_{6\text{-}311}$ method, geometry was reoptimized. The dielectric constant of trifluoroacetic acid used for the solvation calculations is 8.55, molecular size is 4.2Å. All calculations were done using the Firefly v. 7.1.G program [31].

8.3   RESULTS AND DISCUSSION

8.3.1   FORMATION OF TETRAHYDROFURANIUM CATIONS BY CYCLIZATION OF UNSATURATED ALKOXYCARBENIUM IONS

As noted earlier, in the case of the first mechanism of hydrogenated furan forming reaction proceeding in trifluoroacetic acid medium, the rate-controlling step is considered to be the processes of intramolecular cyclization of unsaturated alkoxycarbenium ions. Interest in these reactions is caused also due to the methods for the synthesis of heterocyclic compounds (hydrogenated pyrans and furans) where these compounds take part as intermediates have attracted great attention from researchers lately.

Theoretical values of the thermodynamic parameters for alkyl substituted allyloxymethyl cation cyclizations indicate that the introduction of the substituent into the cation, irrespective of position, promotes formation of hydrogenated furans (Fig. 8.1). So, the Gibbs free energy value of the nonsubstituted ion cyclization is 21.6 kJ/mol whereas reactions with γ-alkyl substituted cations 2–5 proceed with ΔG° values equal to −15–30 kJ/mol.

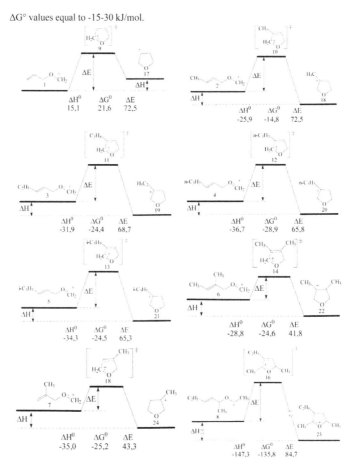

FIGURE 8.1 The potential energy surfaces of nonsaturated alkoxycarbenium ions cyclization reactions (DH^0_{gas}, DG^0_{gas}, DE_{gas} in kJ/mol).

The Gibbs free energies for the cyclizations of the β-substituted (7), β,γ-disubstituted (6) and γ-substituted alkoxycarbenium cations are nearly equal, which is evidence of the similar favorability of the indicated reactions.

The presence of the methyl substituents in the α- and α'-positions of the γ-ethyl-substituted allyloxymethyl cation 8 leads to a Gibbs free energy of cyclization which is more than 4 times lower. This effect is explained by rearrangements of produced tetrahydrofuranium cations into more stable ones. In the case of cyclization β-and γ- substituted allyloxymethyl ions 1–7, the tertiary tetrahydrofuranium cations with the positive charge in the β-position form while α and α'-substituted ions are transformed into the oxonium tetrahydrofuranium cations where the positive charge is stabilized by the unshared electron pair of the oxygen atom.

The presence of alkyl substituents on the double bond of allyloxymethyl cation motivates a decrease of value of activation energy of cyclization.

Cyclization of γ-alkyl substituted ions in comparison with nonsubstituted ion 1, with the exception of methyl substituted ion, is characterized by slightly smaller values of ΔE_{gas}. The cations 6–7 substituted in β – position cyclize with the activation energy, which is 30 kJ/mol lower then the ΔE_{gas} for the γ – substituted ions. Introduction of a second alkyl substituent into the γ-position of β-alkyl-substituted ion (6) does not affect the value of activation energy. The transformation of the α,α'-disubstituted ion 8 in comparison with nonsubstituted ion is characterized by the high value of the activation energy. It is explained by steric obstacles during the cyclization created by the substituent at the α'-position.

It is important to point out that the low values of activation energy observed for the β-alkyl substituted allyloxymethyl ions with the rather high activation energy for cyclization of the γ-alkyl substituted allyloxymethyl ions contradict experimental data, since the nonterminal alkenes have not been involved in the reaction.

8.3.2 STUDY OF THE RECYCLIZATION OF THE DIOXOLENIUM IONS

As has been shown in the previous section, the rate-controlling step of the hydrogenated furan forming reaction occurring in trifluoroacetic acid medium can proceed with participation of the both terminal and nonterminal alkenes when the mechanism that includes the unsaturated alkoxycarbenium ions as intermediates is realized. As this conclusion doesn't explain the reason for involving only terminal alkenes into the reaction, we support and investigate the alternative mechanism for the studied reaction including the step of recyclization of the dioxolenium ions.

The value of Gibbs free energy of the nonsubstituted dioxolenium ion 24 cyclization equals −11.1 kJ/mol and shows that this reaction is a spontaneous process (Fig. 8.2). The substituent at the 1st carbon atom in the hydroxyethyl radical on the dioxolenium ion raises the thermodynamic advantage of the hydrogenated furans formation: in the case of the methyl substituent the $\Delta G°_{gas}$ is −15.3kJ/mol, the ethyl group (26) reduces the value of the Gibbs free energy to −21.4 kJ/mol.

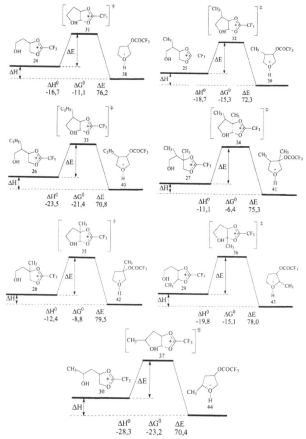

FIGURE 8.2 The potential energy surfaces of dioxolenium ions recyclization reactions ($DH°_{gas}$, $DG°_{gas}$, DE_{gas} in kJ/mol).

The substituent at the 3rd position, in contrast, diminishes the thermodynamic advantage of the recyclization. The $\Delta G°_{gas}$ value for the reaction of the monoalkyl substituted ion 28 is −8.8 kJ/mol. The introduction of the second methyl groups at the 1st carbon atom in the hydroxyethyl substituent of the ion 28 (27) leads to the highest value of the $\Delta G°_{gas}$ among calculated data (−6.4 kJ/mol). It indicates the small thermodynamic advantage of the recyclization of the ions formed from nonterminal alkenes.

In comparison with the cyclization of the nonsubstituted 24, the cyclization of the alkyl substituted dioxolenium ions (29, 30), which can form as a result of the interface reactions with participation of formaldehyde and acetaldehyde, are characterized by lower Gibbs free energy. In case of the ion 30 transformation this value is −23.2 kJ/mol, if there is cyclization of the ion 29 substituted in the 4 position $\Delta G°_{gas}$ is −15.1 kJ/mol.

The activation energy of the nonsubstituted ion recyclization 24 is 76.2 kJ/mol. Methyl groups in the 1st carbon atom in the hydroxyethyl substituent favors the cyclization by reducing the activation energy to 72.3 kJ/mol. In case of the ethyl substituent a further, but not so noticeable, decrease in this value is observed (70.8 kJ/mol).

Research has shown that the dioxolenium ions which are monoalkyl-substituted in the 3 position cyclize with higher energy barriers than the unsubstituted ion. So the activation energy of the methyl substituted ion cyclization 28 is 79.5 kJ/mol. In the presence of the additional methyl group in the 1st position of the hydroxyethyl substituent (27), the activation energy falls to 75.5 kJ/mol.

Because of the steric factors, the methyl substituent in the 4 position of the dioxolenium ring (29) causes an increase in the activation energy value to 78 kJ/mol. On the contrary, the ion 30 substituted at the second carbon atom of the hydroxyethyl substituent is characterized by the lowest activation energy, which is equal to 70.4 kJ/mol.

On the basis of the obtained results, it is possible to come to a conclusion that the ions formed from the terminal alkenes are most easily involved in the recyclization reaction leading to formation of the furan structures. Ions formed from the nonterminal alkenes possess the least propensity for cyclization since the thermodynamic advantage of their transformation is much lower and the activation energy is appreciably higher then the energetic parameters of the reactions including corresponding ions made with the participation of the terminal alkenes. This result matches with the experimental data.

8.3.3 A COMPARISON OF THE REACTION MECHANISMS

To establish the most probable mechanism for formation of hydrogenated furans under Prins reaction conditions, the total change of the Gibbs free energy ($\Delta_r G_{298}$) of generation of the allylic alcohol and the allylic alcohol ester, being precursors of the unsaturated alkoxycarbenium and the dioxolenium ions, has also been calculated.

$$CH_3\text{-}CH=CH_2 + CH_2O \longrightarrow CH_3\text{-}CH=CH\text{-}CH_2\text{-}OH \qquad \text{-124 kJ/mol}$$
$$CH_3\text{-}CH=CH_2 + CH_2O + CF_3COOH \longrightarrow CH_3\text{-}CH=CH\text{-}CH_2\text{-}OCOCF_3 + H_2O \quad \text{-178 kJ/mol}$$

According to these values, under the conditions of the studied reaction the possible competitive formation of the allylic alcohols and the allylic alcohol esters, from thermodynamic points of view, should lead to the esters being primary. One more argument in favor of the second mechanism is that the form of formaldehyde in trifluoroacetic acid medium is mono(trifluoroacetate) of methylenglycol.

Comparison of the Gibbs free energies for cyclizations of ions formed from terminal alkenes (Table 8.3) demonstrates a similar thermodynamic advantage of the cyclization both the unsaturated alkoxycarbenium and dioxolenium ions. The difference between the Gibbs free energies of two reactions is only 3 kJ/mol. The energy barrier is higher for the unsaturated alkoxycarbenium ions cyclization.

TABLE 8.3 Energetics of Intramolecular Cyclization of Unsaturated Alkoxycarbenium Ions and Recyclization of Substituted Dioxolenium Ions into the Tetrahydrofuranium Cations in the Gas Phase (ΔG^0_{gas}, $\Delta G^\#_{gas}$, kJ/mol) and in Solution (ΔG^0_{sol}, $\Delta G^\#_{sol}$, kJ/mol).

Compounds	DG^0_{gas}	$DG^\#_{gas}$	DG^0_{sol}	$DG^\#_{sol}$
C_2H_5 ... $O\overset{+}{=}CH_2$ **3**	−24.4	76.7	−48.6	92.8
CH_3 ... CH_3 ... $O\overset{+}{=}CH_2$ **6**	−24.6	49.4	−17.1	87.3

TABLE 8.3 *(Continued)*

Compounds	DG^0_{gas}	$DG^{\#}_{gas}$	DG^0_{sol}	$DG^{\#}_{sol}$
C₂H₅ structure 26	-21.4	71.1	-27.0	83.1
CH₃ CH₃ structure 27	-6.4	77.6	-5.1	92.1

Alkyl substituents in the β-position of the allyloxymethyl ions and in the 3 position of the dioxolenium ions, which are characteristic for ions generated from the nonterminal alkenes, has exactly the opposite influence on the studied reactions. Gibbs energy of activation obtained for cyclization of the β-substituted unsaturated alkoxycarbenium ions (5) is lower than Gibbs energy of activation for the γ-alkyl substituted ions (3) by 25.0 kJ/mol. The thermodynamic advantage for both types of the alkoxycarbenium ions is approximately the same. In a case of the dioxolenium ions, the substituent in the 3 position raises values both the ΔG°_{gas} and the $\Delta G^{\#}_{gas}$. So, the comparison of the reactions of the isomer dioxolenium ions 26 and 27 showed that the ion 27 substituted in three-position cyclize with Gibbs free energy to 15kJ/mol and Gibbs energy of activation to 5kJ/mol higher than the ion 26.

The Gibbs free energies and the Gibbs free energies of activation of formation of tetrahydrofuranium cations 19, 22, 40, 41 calculated from the solvation energies are in a reasonable agreement with the corresponding gas-phase energies (Table 8.3). Consideration of the solvent influence does not change the free energy of Gibbs of recyclization of the ions 26, 27 very much, the Gibbs energies of activation for these two reactions are increased approximately 10-15 kJ/mol. The reaction barrier for the dioxolenium ion 26 formed from the terminal alkene, as well as in case of calculations for the gas phase, by 10 kJ/mol lower than for recyclization of the ion 27.

A study of cyclization of the unsaturated alkoxycarbenium ions 3 and 6 with taking into account effects of solvent also confirms the results received by gas phase calculation: the cyclization of the β-substituted ions is more favorable. At the same time the height of the reaction barrier for the ion 6 is about 40 kJ/mol larger than the values obtained earlier. The Gibbs free energy of activation of the γ-substituted allyloxymethyl ion cyclization calculated with taking into account effects of solvent is equal 92,8kJ/mol that mean that γ-substituted ions seems to be the most unreactive from the kinetic point of view among considered. Gibbs›s free energy of the cyclization of γ-substituted ion determined using the PCM method diminished to −48kJ/mol.

The comparison of the mechanisms demonstrated a greater kinetic advantage of the cyclization of the unsaturated alkoxycarbenium and the dioxolenium ions formed from the terminal alkenes. Participation in the reaction of nonterminal alkenes reduces the advantage of the dioxolenium ion cyclization, which agrees with experiment.

8.4   CONCLUSION

1. By means of the quantum-chemical modeling using compound methods reproducing results for MP4(full)/6-311G+(fd,p) approach, thermodynamic and kinetic parameters of the two possible mechanisms of the hydrogenated furans formation under Prins reaction conditions in trifluoroacetic acid medium are determined.

2. According to calculated values of the Gibbs free energy, transformations of the allyloxymethyl and 3-(2-hydroxyethyl)-1-trifluoromethyl-2,5-dioxolenium cations into the hydrogenated furans are spontaneous processes. However, the trifluoromethyl-dioxolenium ion is preferably formed in trifluoroacetic acid from these specified intermediates.

3. The determined dependence of the Gibbs free energy and the activation energy of recyclization of the 3 (2-hydroxyethyl)-1-trifluoromethyl-2,5-dioxolenium ions on their structure coincides with experimentally observed activity of the alkenes in trifluoroacetic acid in the 3-alkylsubstituted hydrogenated furans formation reaction. According to this dependence, the recyclization of the dioxolenium ions formed from nonterminal alkenes is characterized by

greater values of the Gibbs free energy and the activation energy. The dependence, which was defined for transformations of the allyloxymethyl cations predicts that nonterminal alkenes should be more reactive in this reaction.

4. It is theoretically proved that the mechanism of formation of 3-alkylsubstituted hydrogenated furans in trifluoroacetic acid by interaction between terminal alkenes and formaldehyde medium which includes recyclization of 3-(2-hydroxyethyl)-1-trifluoromethyl-2,5-dioxolenium ions is preferable.

KEYWORDS

- **Hydrogenated Furans**
- **The Prins Reaction**
- **Trifluoroacetic Acid**

REFERENCES

1. Talipov, R. F., Sagitdinova, K. F., Vakulin, I. V., Safarov, M. G. (2000). Syntheses Based on 3-Chloroacetoxytetrahydrofuran, Russ, J. Org. 36, 1198–1200.
2. Talipov, R. F., Safarov, I. M., Talipova, G. R., & Safarov, M. G. (1997). Kinetics of Heptene-I Interaction with Formaldehyde in Trifluoroacetic Acid. Reaction Kinetics and Catalysis Letters, 61, 63–68.
3. Talipov, R. F., Safarov, M. G. (1997). Prins Reaction AdE as a Set of Transformations United by a Common Name. Bashkir J. Chem., 4, 10–13.
4. Talipov, R. F., Muslukhov, R. R., Safarov, I. M., Yamantaev, F. A., Safarov, M. G. (1996). Synthesis of β-Substituted Tetrahydrofurans by Prins Reaction, ChemInform, doi: 10.1002/chin.199604138.
5. Shepelevich, I. S, Talipov, R. F. (2002). RHF-MNDO Calculation of the Enthalpy of Formation for Substituted 1-hydroxy-2-oxa-5-pentyl Carbocations, J Struct Chem, 43, 858–861.
6. Talipov, R. F., Starikov, A. S., Gorina, I. A., Akmanova, N. A., Safarov, M. G. (1993). Synthesis of Dihydrofurans by Prince Reactions in the Trifluoroacetic-acid Medium, Russ, J. Org. 29, 1024–1027.
7. Talipov, R. F., Starikov, A. S., Gorin, A. V., & Safarov, M. G. (1993). 1-Step Synthesis of 2, 3, 5-Trialkyl-2, 5-Dihydrofurans Using the Prince Reaction, Russ J Org 29, 748–750.
8. Talipov, R. F., Mustafin, M. M., & Safarov, M. G. (1993). Prince Reaction with Participation of 4-vinyl-1-cyclohexen, Russ J. Org. 29, 127–129.

9. Greenwood, J. R., Capper, H. R., Allan, D. R., & Johnston G. A. R. (1997). Tautomerism of Hydroxy-pyridazines, The N-oxides, Theochem, 419, 97–111.
10. Yang, X., Mague, J. T., & Li, C. (2001). Diastereoselective Synthesis of Polysubstituted Tetrahydropyrans and Thiacyclohexanes via Indium Trichloride Mediated Cyclizations, J. Org Chem, 66, 739–747.
11. Loh, T., Hu, Q., & Ma, L. (2001). Formation of Tetrahydrofuran From Homoallylic Alcohol via a Tandem Sequence, 2-oxonia [3, 3] Sigmatropic Rearrangement/Cyclization Catalyzed by In(OTf) 3, J. Am. Chem. Soc., 123, 2450–2451.
12. Suginome, M., Iwanami, T., & Ito, Y. (1998). Stereoselective Cyclization of Highly Enantioenriched Allylsilanes with Aldehydes via Acetal Formation, New Asymmetric Access to Tetrahydropyrans and Piperidines, J., Org Chem 63, 6096–6097.
13. Cohen, F., MacMillan, D. W. C., Overman, L. E., & Romero, A. (2001). Stereoselection in the Prins-pinacol Synthesis of Acyltetrahydrofurans, Organic Letters, 3, 1225–1228.
14. Hanaki, N., Link, J. T., MacMillan, D. W. C., Overman, L. E., Trankle, W. G., & Wurster, J. A. (2000). Stereoselection in the Prins-pinacol Synthesis of 2, 2-Disubstituted 4-Acyltetrahydrofurans, Enantioselective Synthesis of (-) Citreoviral, Organic Letters, 2, 223–226.
15. Overman, L. E., Pennington, L. D. (2003). Strategic Use of Pinacol-Terminated Prins Cyclizations in Target-oriented total synthesis. J., Org Chem 68, 7143- 7157.
16. Jaber, J. J., Mitsui, K., Rychnovsky, S. D. (2001). Stereoselectivity and regioselectivity in the segment-coupling Prins cyclization. J. Org. Chem. 66, 4679–4686.
17. Miles, R. B., Davis, C. E., & Coates, R. M. (2006). Syn- and anti-selective Prins cyclizations of δ,ε-unsaturated ketones to 1,3-halohydrins with Lewis acids. J Org Chem 71, 1493–1501.
18. Jasti, R., & Rychnovsky, S. D. (2006). Racemization in Prins cyclization reactions. J., Am Chem Soc 128, 13640–13648.
19. Larock, R. C., Hightower, T. R., Hasvold, L. A., & Peterson, K. P. (1996). Palladium(II)-catalyzed cyclization of olefinic tosylamides, J Org Chem 61, 3584–3585.
20. Huang, Q., & Larock, R. C. (2003). Synthesis of 4-(1-alkenyl)isoquinolines by palladium (II)- catalyzed cyclization/olefination, J., Org Chem 68, 980–988.
21. Roesch, K. R., & Larock, R. C. (2002). Synthesis of isoquinolines and pyridines by the palladium/copper-catalyzed coupling and cyclization of terminal acetylenes and unsaturated imines. The total synthesis of decumbenine. J. Org Chem 67, 86–94.
22. Qing, F., Gao, W., & Ying, J. (2000). Synthesis of 3-trifluoroethylfurans by palladium-catalyzed cyclization-isomerization of (Z)-2-alkynyl-3-trifluoromethyl allylic alcohols, J Org Chem 65, 2003–2006.
23. Smith, M. B, & March, J. (2001). March's advanced organic chemistry. Reactions, mechanisms, and structure (fifth edition). John Wiley & Sons, New York.
24. Curtiss, L. A., Raghavachari, K., Redfern, P. C., Rassolov, V., & Pople, J. A. (1998). Gaussian-3 (G3) theory for molecules containing first and second-row atoms J Chem Phys 109, 7764–7777.
25. Curtiss, L. A., Raghavachari, K., Redfern, P. C., Pople J. A. (1997). Assessment of Gaussian-2 and Density Functional Methods for the Computation of Enthalpies of Formation. J Chem Phys 106, 1063.

26. DeYonker, N. J., Cundari, T. R., & Wilson, A. K. (2006). The correlation consistent composite approach "ccCA": An alternative to the Gaussian-n methods. J Chem Phys 124, 114104.

27. Olivella, S., & Sole, A. (2000). Ab Initio Calculations on the 5-exo Versus 6-endo Cyclization of 1, 3-Hexadiene-5-yn-1-yl Radical Formation of the first Aromatic Ring in Hydrocarbon Combustion, J., Am Chem Soc, 122, 11416–11422.

28. Sullivan, M. B., Iron, M. A., Redfern, P. C., Martin, J. M. L., Curtiss, L. A., & Radom, L. (2003). Heats of Formation of Alkali Metal and Alkaline Earth Metal Oxides and Hydroxides, Surprisingly Demanding Targets for High-Level Ab Initio Procedures, J., Phys Chem, A., 107, 5617–5630.

29. Woodcock, H. L., Moran, D., Pastor, R. W., MacKerell, A. D., & Brooks, B. R. (2007). Ab Initio Modeling of Glycosyl Torsions and Anomeric Effects in a Model Carbohydrate, 2-Ethoxy Tetrahydropyran, Biophys, J 93, 1–10.

30. Friesner, R. A., Murphy, R. B., Beachy, M. D., Ringnalda, M. N., Pollard, W. T., Dunietz, B. D., & Cao, Y. (1999). Correlated Ab Initio Electronic Structure Calculations for Large Molecules, J Phys Chem, A., 103, 1913–1928.

31. Granovsky, A. A. http://classic.chem.msu.su/gran/gamess/index.html.

CHAPTER 9

THEORETICAL INVESTIGATION OF 1,3-DIOXANES FORMATION IN PRINS REACTION BY ADDITION OF FORMALDEHYDE OLIGOMERS TO ALKENES

I. V. VAKULIN, R. F. TALIPOV, O. YU. KUPOVA, G. R. TALIPOVA, A. V. ALLAGULOVA, and R. R. FAIZULLINA

CONTENTS

ABSTRACT

Addition of formaldehyde dimer to alkenes with in the gas phase by the Prins reaction are studied at the MP2(fc)/6-31G(d,p). It is shown that the initially complex obtained at the first study turn into 1,3-dioxane without intermediate formation of a σ-complex. Also shown, that transformation of π-cation is a pseudo synchronous process.

9.1   INTRODUCTION

The Prins reaction is a known method of O-containing heterocycles formation [1, 2]. In some cases the insufficient selectivity of this reaction is a drawback. For instance, the first stage of the isoprene synthesis by the "dioxane" method [3] is accompanied by the formation of a large number of methyldihydropyrans [4–6]. So, investigation and specification of mechanism of Prins reaction can solve this problem.

The generally accepted mechanism of the Prins reaction presented on Fig. 9.1.

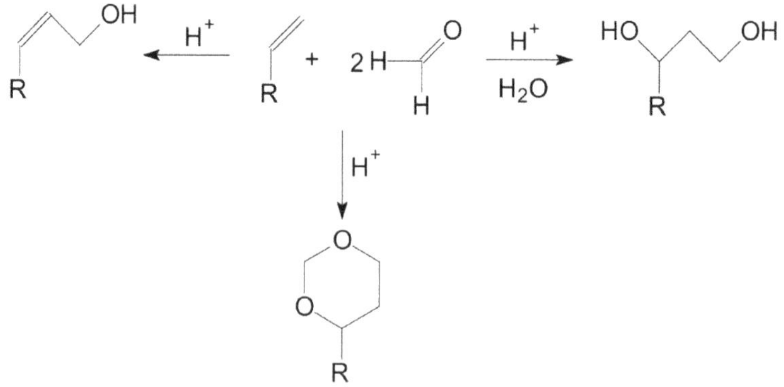

FIGURE 9.1   The Prins reaction products [7, 8].

Experimental [9] and theoretical data [10], [11] clearly demonstrate that the presence of formaldehyde oligomers (FO) is a prerequisite for the 4-alkyl-1,3-dioxanes formation, whereas FO are much more reactive than the monomer. It is assumed that the FO addition at the double bond can be described as sequential, or pseudo synchronous [12] (Fig. 9.2).

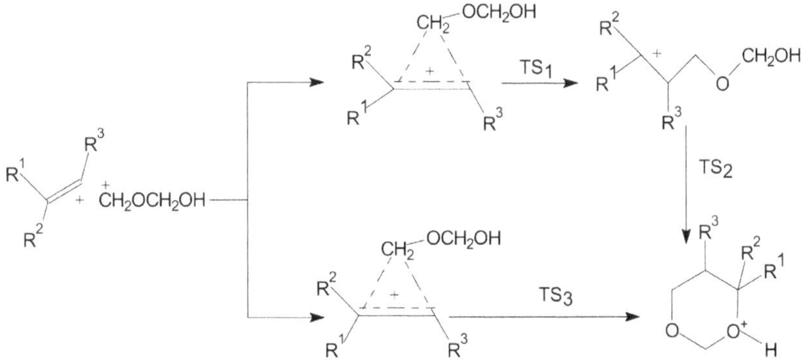

Fig.2. Possible schemes of dimer formaldehydes and alkenes addition, where $R^1=R^2=R^3=H$ (1); $R^1=CH_3$, $R^2=R^3=H$ (2); $R^1=C_2H_5$, $R^2=R^3=H$ (3); $R^1=R^2=CH_3$, $R^3=H$ (4); $R^1=H$, $R^2=R^3=CH_3$ (5).

FIGURE 9.2 Possible schemes of dimer formaldehydes and alkenes addition, where $R^1=R^2=R^3=H$ (1); $R^1=CH_3$, $R^2=R^3=H$ (2); $R^1=C_2H_5$, $R^2=R^3=H$ (3); $R^1=R^2=CH_3$, $R^3=H$ (4); $R^1=H$, $R^2=R^3=CH_3$ (5).

The implementation of the two mechanisms corresponds to the experimental data on a stereoselective Prins reaction for a series of cycloalkenes [13].

In the work we studied some features of formaldehyde oligomers and alkenes interaction resulting in the formation of 1,3-dioxanes. The geometric and orbital structures of the reactions transition states depending on the alkene structure are considered. The following alkenes were used: ethylene (1), propylene (2), butene 1 (3), isobutylene (4) and t-butene-2 (5). A formaldehyde dimer (FD) served an example for FO calculation.

9.2 METHODS OF CALCULATIONS

PC GAMESS Firefly v7.1 programs [14] were used in carrying out quantum and chemical calculations. Searching for the equilibrium geometry of the transition states was carried out by MP2(fc)/6-31G(d) [15–17] approximation. Verification of the transition state was made by calculating vibrational parameters for the obtained transition state geometry with the subsequent analysis of the calculated frequencies of the IR spectrum, as well as modeling the found transition state into the starting materials and final products by means of the IRC procedure. The degree of asynchrony in the cyclization reaction was calculated by the method [18].

9.3 CALCULATION RESULTS AND DISCUSSION

Due to the calculation results, the section of the potential energy surface corresponding to the formation of 4-alkyl-1,3dioxane from FD and al-kenes (1–5) is presented as follows (Fig. 9.3):

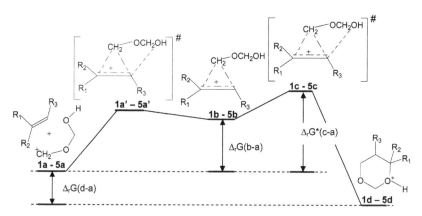

FIGURE 9.3   Potential energy surface of the Prins reaction.

It is shown that this reaction proceeds in two stages without an inter-mediate σ-cation formation. The π-cation (1b–5b) is directly transformed into the 1,3-dioxane structure (1d–5d) though the corresponding transition states (1c–5c). The calculated values of the thermodynamic parameters are presented in Table 9.1.

TABLE 9.1   Energy Parameters of Dioxane Formation Reactions, kJ/mol.

Alkene	$\Delta_r G^*(a–c)$	$\Delta_r G(b–a)$	$\Delta_r G(d–a)$	$\Delta_r G^*(c–b)$
1	138.7	110.9	−7.4	27.7
2	99.2	92.6	−27.6	6.6
3	72.2	87.9	−30.5	18.5
4	66.2	36.9	−33.8	29.3
5	114.7	67.8	−35.3	46.9

According to the Gibbs free energy values, a dioxane (1d–5d) formation is thermodynamically favorable; it slightly increases while the number of alkyl substituents at the double bond grows. Thus, the Gibbs free energy value for ethylene amounts to −7.4 kJ/mol, whereas this value for butene-2 is −35.3 kJ/mol.

The calculated values of the Gibbs activation energy reaction show that ethylene is hard to be reacted ($\Delta rG^*(a–d) = 138.7$ kJ/mol). The lowest values of the activation energy are achieved in reacting with terminal alkenes, propylene and butane-1 and amount to 99.2 and 72.2 kJ/mol accordingly. In increasing the number of alkyl substituents at the double alkene bond, the Gibbs activation energy slightly grows (for butane-2 and isobutylene). Such character of changes in the activation energy of the alkene structure is well explained by a steric factor.

Besides, such potential energy surface points to the fact that the limiting stage of the alkene reacting with FD is the formation of π-cation.

The structure of transition states (1c–5c) indicating the values of interatomic distances and bond orders of the reaction are presented in Fig. 9.1.

According to the data presented in almost all transition states the formation of the C–C bond of 1,3 dioxane cycle is accomplished earlier than the O–C bond one. So the C2–C4 in the transition states 1c, 2c, 4c, 5c fall within 1.811 till 2.286 Å, whereas the order of the bond is no less than 0.244. At the same time in transition states 2c–5c interatomic distance O1–C3 corresponding to the other bond of the 1,3-dioxane cycle ranges from 2.526 to 3.124 Å with the bond order 0. It proves the consecutive formation of bonds. The exception lies in the transition states 1c for the reaction of ethylene and FD. In this case the O1–C3 distance amounts to 2.526 Å and the bond order is 0.07.

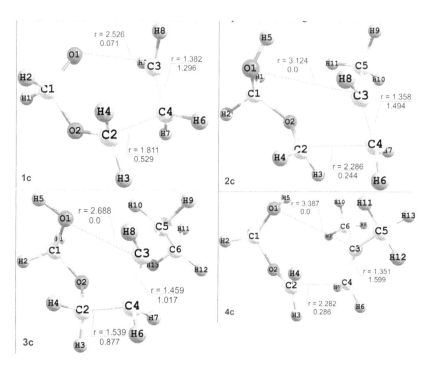

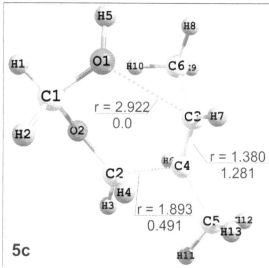

FIGURE 9.4 Structure of transition states (1c–5c).

The structure of the 3c transition state for butane-1 and FD reaction somehow differs from the structure of other transition states. It can be considered as an intermolecular cyclization of σ-cation. In fact the distance and bond order C2–C4 values are characteristic for a single bond C–C and amount to 1.539 and 0.877 Å accordingly. The O3–C1 bond has not been formed (R = 2.688 Å) and the distance and bond order values C3=C4 amount to 1.459 Å and 1.017. And this bond is quite similar to the single bond C–C at C+ in carbocation. These parameters considerably differ from the similar values of other transition states where the C3–C4 bond can be treated as a sesquialteral one. However, a stable σ-cation in the butane-1 and FD reaction was not discovered. The forming π-cation 3b as in other alkenes is directly transformed into a dioxane structure.

An orbital structure, population and energy of bordering orbitals of the transition states 1c–5c are presented in Fig. 9.5 and Table 9.2.

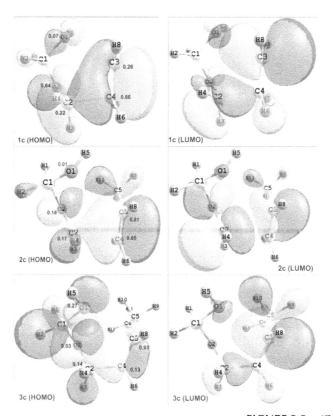

FIGURE 9.5   *(Continued)*

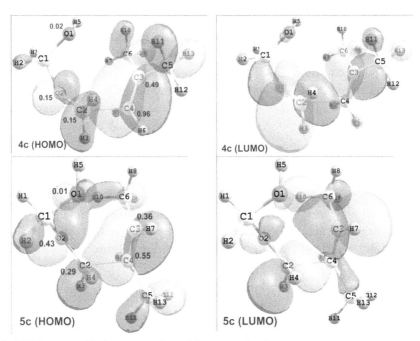

FIGURE 9.5   Orbital structure of transition states (1c–5c).

TABLE 9.2   Activation Energy of Forming 4-alkyl-1,3-Dioxanes and Some Peculiarities of Electronic Structure of the Corresponding TS (1c–5c).

TS	$\Delta_r G^{\neq}_{298}$, kJ/mol	E_{HOMO}, eV	E_{LUMO}, eV	IP, eV
1c	138.7	−16.29	−4.01	12.28
2c	99.2	−14.92	−2.40	12.51
3c	72.2	−16.24	−3.72	12.52
4c	66.2	−14.25	−2.67	11.58
5c	114.7	−15.49	−2.64	12.85

According to the data presented, the value of the ionization potential (IP) for the transition states is correlated with the activation energy and increases as the number of alkyl substituents at the double bond grows. For ethylene of the transition state 1c the IP value amounts to 12.28 eV whereas this value of butane-2 is already 12.28 eV. The exception is the isobutylene the IP and $\Delta_r G{\neq}298$ values of which are minimum in the considered alkene series. It is probably explained by I^+ effect of two methyl substituents leading to a considerate distortion in electronic density of the double bond.

9.4   CONCLUSION

Due to the analysis of the calculated data of the transition states structure of 4-alkyl-1,3-dioxane formation from formaldehyde oligomers and alkenes, it is found that 1,3-dioxane structures are formed in the result of direct isomerization of π-cation on the first stage. A free σ-cation formation is not observed here.

KEYWORDS

- **1.3-Dioxanes**
- **AB Initio Calculation**
- **Alkenes**
- **Formaldehyde Oligomers**
- **Transition State**

REFERENCES

1. Carballo, R. M., Ramirez, M. A., Rodriguez, M. L., Martin, V. S., & Padron, J. I. (2006). Iron (Ш)-Promoted Aza-Prins-Cyclization, Direct Synthesis of Six-Membered Azacycles, Org. Lett. 8, 3837–3840.
2. Bach, T., & Löbel, J. (2002). Selective Prins Reaction of Styrenes and Formaldehyde Catalyzed by 2, 6-Di-tert-butylphenoxy (difluoro) borane, Synthesis 172521–2526.
3. Bogatyrev, V. F., Publ. 27.07.08 Patent RU 2 330 010 C2. Way of Isoprene Obtaining, 21.
4. Chavre, H., Choo, J. K., Lee, A. N., Pac, Y., & Kim, Y. S. (2008). Cho, 5- and 6-Exocyclic Products, cis-2,3,5-Trisubstituted Tetrahydrofurans, and cis-2, 3, 6-Trisubstituted Tetrahydropyrans via Prins-Type Cyclization, J. Org. Chem. 73, 7467–7471.
5. Yadav, J. S., Reddy, B. V. S., Kumar, G. G. K. S. N., & Aravind, S. (2008). The Aqueous Prins-Cyclization, A Diastereoselective Synthesis of 4-hydroxytetrahydropyran Derivatives, Synthesis 3, 395–400.
6. Aubele, A. L., Lee, C. A., & Floreancig, P. E. (2003). The "aqueous" Prins reaction, Org. Lett. 23 4521–4523.
7. Tian, G-Q., & Shi, M. (2007). 2-(Arylmethylene)cyclopropylcarbinols could be converted to the corresponding tetrahydropyrans stereoselectively in the presence of Brønsted acids under mild conditions. A plausible Prins-type reaction mechanism has been proposed, Org. Lett. 9 2405–2408.

8. Isidro, M., & Pastor, Miguel Yus. (2007). The Prins Reaction Advances and Applications, Curr. Org. Chem. 10, 925–957.
9. Talipov, R. F. (1984). Dihydro- and alkeniltetrahydropyrans in the Prins Reaction. Ufa.
10. Vakulin, I. V. (2002). Research of heterocycle formation by the Prins reaction and their other transformations Moscow.
11. Shepelevich, I. S., & Talipov, R. F. (2003). Cyclization of cyclohexene with formaldehyde in trifluoroacetic acid, Bashkir.Khim J., 10 58–60.
12. Overman, L. E., & Pennington, L. D. (2003). Strategic Use of Pinacol-Terminated Prins Cyclizations in Target-Oriented Total Synthesis, J. Org. Chem. 68, 7143–7157.
13. Taber, D. F. (2006). Stereoselective Construction of Oxygen Heterocycles, Org. Chem. Highlights.
14. Granovsky, A. A. http://classic.chem.msu/gran/gamess/index.html.
15. Hehre, W. J., Radom, L., Schleyer, P. R., & Pople, J. A. (1986). Ab Initio Molecular Orbital Theory, Wiley, New York.
16. Freeman, F., Kasner, M. L., & Hehre, W. J. (2001). An Ab initio molecular orbital theory study of the conformational free energies of 2-methyl-, 3-methyl-, and 4-methyltetrahydro-2H-pyran, Journal of Molecular Structure, THEOCHEM 574, 19–26.
17. Kestutis, A., Mikkelsen, K. V., & Stephan, P. A. (2008). Sauer On the Accuracy of Density Functional Theory to Predict Shifts in Nuclear Magnetic Resonance Shielding Constants due to Hydrogen Bonding, J. Chem. Theory Comput. 4, 267–277.
18. Castillo, R., Moliner, V., & Andrés, J. (2000). A theoretical study on the molecular mechanism for the normal Reimer–Tiemann reaction, Chem. P. Lett. 318, 270–275.

CHAPTER 10

QUANTUM-CHEMISTRY STUDY OF THE FEATURES OF PARTICIPATION OF FORMALDEHYDE OLIGOMERS IN PRINS REACTION

I. V. VAKULIN, R. F. TALIPOV, O. YU. KUPOVA, G. R. TALIPOVA, A. V. ALLAGULOVA, and R. R. FAIZULLINA

CONTENTS

ABSTRACT

Nature of addition of formaldehyde to alkenes in Prins reaction is studied by MP2(fc)/6-31G(d). The structure of the transition states and intermediates are found. It is shown that the hydrogenated pyrans as 1,3-dioxanes can be formed by addition of formaldehyde oligomers to alkenes. However, the activation energy of this reaction is higher than that of 1,3-dioxanes formation.

10.1 INTRODUCTION

The Prins reaction is a one of example of multichannel reaction. This reaction became basis for a convenient technique of oxygen-containing heterocycles formation [1, 2]. In some cases this multipathing is drawback because of insufficient selectivity. For example, the first stage of the isoprene synthesis by the "dioxane" method [3] is accompanied by the formation of a large number of by-methyldihydropyrans [4–6]. Obviously, in order to find new way to increase the selectivity of the first stage the mechanism of Prins reaction should be improved. Products formation by way of cascade involvement of one or two molecules of formaldehyde monomer [7, 8] is considered one of the generally accepted mechanisms of the Prins reaction (Fig. 10.1).

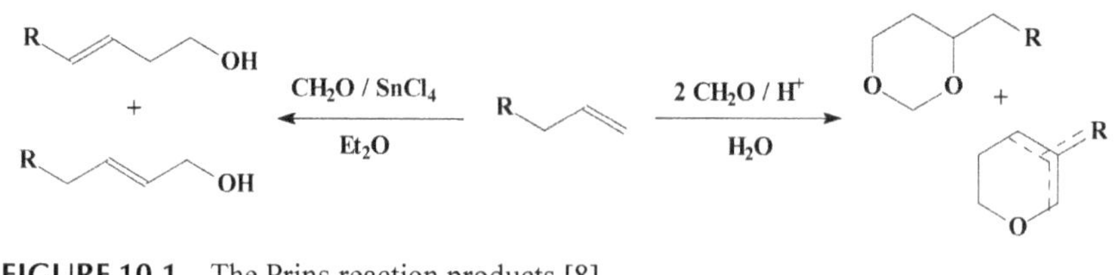

FIGURE 10.1 The Prins reaction products [8].

Selectivity of this reaction can be defined not only by structure of alkenes, but formaldehyde monomer to oligomers ratio too. Experimental [9] and theoretical data [10], clearly demonstrate that the presence of formaldehyde oligomers (FO) is a prerequisite for the 4-alkyl-1,3-dioxanes formation, whereas FO are much more reactive than the monomer. It is shown that FO can play important role in formation of dihydropyrans from alkenes with endo-double bond in non-aqueous solvents [11].

In water dihydropirans become a major product of Prins reaction only in interaction of formaldehyde monomer with alkenes with exo-double bond [12,13].

It is assumed that the FO addition at the double bond can be described as sequential, or pseudo synchronous [14].

The implementation of the two mechanisms corresponds to the experimental data on a stereoselective Prins reaction for a series of cycloalkenes [15]. However, the role of formaldehyde oligomers in the Prins reaction and its reactivity in addition to the alkenes with various structures are still poorly described.

In this work we studied some features of formaldehyde oligomers and alkenes interaction resulting in the formation of oxygen-containing heterocycles. The geometric structures of the key intermediates and reactions transition states are considered depending on the alkene structure. The following alkenes were used as model compounds: ethylene (1), propylene (2), 1-butene (3), isobutylene (4) and *trans*-2-butene (5). A formaldehyde dimer (FD) served an example for FO calculation.

10.2   METHODS OF CALCULATIONS

PC GAMESS v7.1 programs [16] were used in carrying out quantum and chemical calculations. Searching for the equilibrium geometry of the transition states was carried out by MP2(fc)/6-31G(d) [17–20] approximation. Verification of the transition state was made by calculating vibrational parameters for the obtained transition state geometry with the subsequent analysis of the calculated frequencies of the IR spectrum, as well as modeling the found transition state into the starting materials and final products by means of the IRC procedure [21]. The degree of asynchrony in the cyclization reaction was calculated by the method [22]. Bond orders were calculated according [23].

10.3   CALCULATION RESULTS AND DISCUSSION

10.3.1.   π-CATION FORMATION

Detailed analysis of the results of the calculations showed that for each of substituted alkenes (2–5) formation of several π-cations [24] is possible.

These π-cations differ by geometry, energy and reactivity. Only ethylene forms one π-cation, which is transformed subsequently into 1,3-dioxane.

For double substituted alkenes, π-cations, which are transformed into 1,3-dioxane(*) or pyrane(**) had different structure and the relative stability. In other case for 1-butene existing reactive π-cation can turn into 1,3-dioxane and pyrane (Fig. 10.2.).

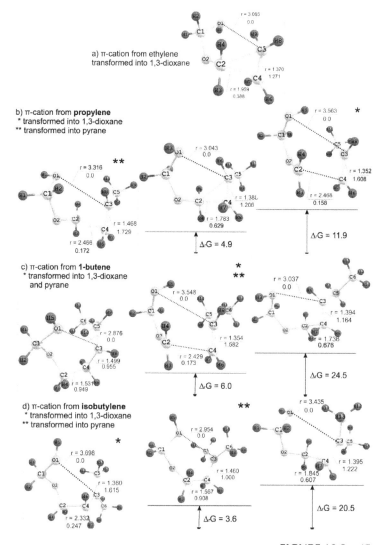

FIGURE 10.2   *(Continued)*

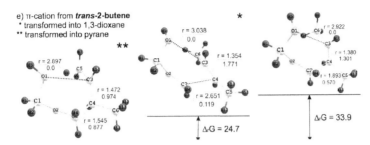

FIGURE 10.2 Geometry and relative stability of π-cations.

10.3.2 1,3-DIOXANES FORMATION FROM FORMALDEHYDE DIMER

Due to the calculation results, the section of the potential energy surface corresponding to the formation of 4-alkyl-1,3dioxane from FD and alkenes (1–5) is presented as follows (Fig. 10.3):

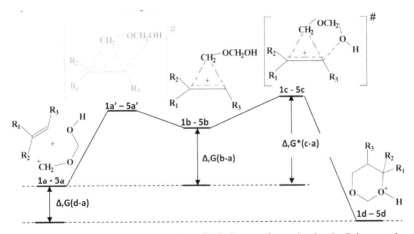

FIGURE 10.3 Potential energy surface of 1,3-dioxane formation by the Prins reaction.

It is shown that this reaction proceeds in two stages without an intermediate σ-cation formation. The π-cation (*1b–5b*) is directly transformed into the 1,3-dioxane structure (*1d–5d*) through the corresponding transition states (*1c–5c*). During the research we were unable to clearly establish the existence of the transition states corresponding to the transformation of reactants into π-cations. We found a number of structures that could be that the transition state, but the IRC procedure does not confirm their reliability. Therefore we do not consider it possible to discuss the unreliable results. The calculated values of the thermodynamic parameters are presented in Table 10.1.

TABLE 10.1   Energy Parameters of Dioxane Formation Reactions, kJ/mol

Alkene	$\Delta_r G(b–a)$	$\Delta_r G(d–a)$	$\Delta_r G^*(c–a)$	$\Delta_r G^*(c–b)$
1	110.9	−7.4	138.7	27.7
2	92.6	−27.6	99.2	6.6
3	47.7	−30.5	72.2	18.5
4	33.2	−33.8	66.2	29.3
5	67.8	−35.3	114.7	46.9

Transition from reagents to p-cation is marked as (**b–a**) and transition from reagents to 1,3-dioxane structure as (**c–a**). $\Delta_r G^*(c–a)$ is activation Gibbs energy of the whole reaction and $\Delta_r G^*(c–b)$ is activation Gibbs energy corresponding to microstage from p-cation to 1,3-dioxane.

According to the Gibbs free energy values, a dioxane (*1d–5d*) formation is thermodynamically favorable; it slightly increases while the number of alkyl substituents at the double bond grows. Thus, the Gibbs free energy value for ethylene amounts to −7.4 kJ/mol, whereas this value for 2-butene is −35.3 kJ/mol.

The calculated values of the Gibbs activation energy reaction show that ethylene is hard to be reacted ($\Delta_r G^*(d–a) = 138.7$ кJ/mol). The lowest values of the activation energy are achieved in reacting with terminal alkenes, propylene and butane-1 and amount to 99.2 and 72.2 kJ/mol accordingly. In increasing the number of alkyl substituents at the double alkene bond, the Gibbs activation energy slightly grows (for butane-2 and isobutylene). Such character of changes in the activation energy of the alkene structure is well explained by a steric factor.

Besides, such potential energy surface points to the fact that the limiting stage of the alkene reacting with FD is the formation of π-cation.

The structure of transition states (*1c–5c*) indicating the values of interatomic distances and bond orders of the reaction are presented in Fig. 10.4.

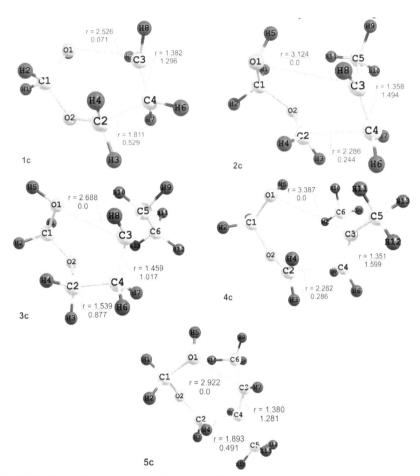

FIGURE 10.4   Structure of transition states (*1c–5c*).

According to the data presented in almost all transition states the formation of the C–C bond of 1,3 dioxane cycle is accomplished earlier than the O–C bond one. So the C2–C4 in the transition states *1c, 2c, 4c, 5c* fall within 1.811 till 2.286 Å, whereas the order of the bond is no less than 0.244. At the same time in transition states *2c–5c* interatomic distance O1–C3 corresponding to the other bond of the 1,3-dioxane cycle ranges from 2.526 to 3.124 Å with the bond order 0. It proves the consecutive formation of bonds. The exception lies in the transition states 1c for the reaction of ethylene and FD. In this case the O1–C3 distance amounts to 2.526 Å and the bond order is 0.07.

The structure of the transition state *3c* for butane-1 and FD reaction somehow differs from the structure of other transition states. It can be considered as an intermolecular cyclization of σ-cation. In fact the distance and bond order C2–C4 values are characteristic for a single bond C–C and amount to 1.539 and 0.877 Å accordingly. The O3–C1 bond has not been formed (R = 2.688 Å) and the distance and bond order values C3=C4 amount to 1.459 Å and 1.017. And this bond is quite similar to the single bond C–C at C+ in carbocation. These parameters considerably differ from the similar values of other transition states where the C3–C4 bond can be treated as a sesquialteral one. However, a stable σ-cation in the butane-1 and FD reaction was not discovered. The forming π-cation *3b* as in other alkenes is directly transformed into a dioxane structure.

The analysis of the IRC procedure results (i.e., Fig. 10.5 IRC plot for isobutylene) allows to estimate the degree of synchrony in adding FD by the double bond. To fulfill it, there were found the structures where the second bond order O1–C3 of 1,3-dioxane cycle becomes quite a appreciable value on the way of transforming transition states (*1c–5c*) in the corresponding 4-alkyl-1,3-dioxanes (*1d–5d*). The synchrony degree was calculated by the method suggested in the article [22]. The values of the parameters considered are presented in Table 10.2.

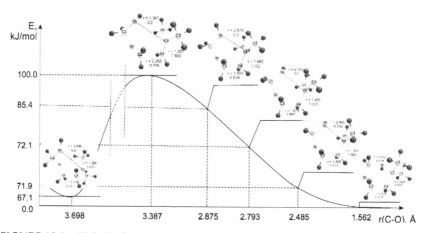

FIGURE 10.5   IRC plot for isobutylene.

TABLE 10.2   Interatomic Distances and Bond Orders for the Structures on the Reaction Coordinate

Alkene	C_2-C_4		O_1-C_3		Synchrony degree, %
	r, A	Bond order	r, A	Bond order	
1*	1.811	0.529	2.526	0.071	40.2
2	1.577	0.890	2.470	0.058	8.7
3	1.540	0.920	2.512	0.057	7.7
4	1.559	0.909	2.485	0.053	9.2
5	1.591	0.884	2.529	0.051	8.6

* – TS.

In all cases except ethylene, the formation of O1–C3 bond occurs with a considerate delay and O1–C3 interaction becomes clear when C2–C4 bond is almost formed (Table 10.2). So the synchrony degree of DF adding by the double bond is estimated by 7.7–9.2%. And only in case of ethylene O1–C3 bond starts forming simultaneously with the C2–C4 bond with the synchrony degree amounting to 40.2%. Such a π-cation transformation results in that fact that dimer formaldehyde addition by the double bond in a gas phase or nonpolar media occurs as a syn-addition.

10.3.3 HYDROGENATED PYRANS FORMATION FROM FORMALDEHYDE DIMERS

According to literature [5], hydrogenated pyrans are thermodynamically more favorable reaction products as compared with 1.3-dioxans. At the same time it is assumed that their formation occurs with the participation of formaldehyde monomer. However, the results of our calculations show that hydrogenated pyrans can also be produced from FO and alkenes (Fig. 10.6).

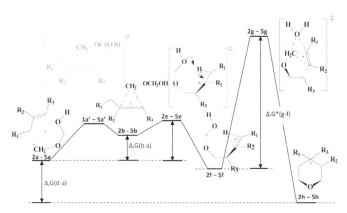

FIGURE 10.6 The scheme of hydrogenated pyrans formation from alkenes and FD. Proton transfer at the stage b→d is suggested to a corresponding mechanism [25].

Together with alkenes (*2a–5a*) FD preliminarily forms the corresponding π-cations (*2b–5b*), which are isomerized through the transition states (*2e–5e*) into σ-cations (*2f–5f*). σ-cations (*2f–5f*) are subsequently transformed via the transition states (*2 g–5 g*) into the pyran cycle (*2h–5h*). The calculated values of thermodynamic parameters of the corresponding reaction stages are presented in Table 10.3.

TABLE 10.3 Energy Parameters of the Hydrogenated Pyrans Formation Reactions, kJ/mol

Alkene	$\Delta_r G^*_{298}$(g–f)	$\Delta_r G^{\neq}_{298}$(e–a)	$\Delta_r G_{298}$(b–a)	$\Delta_r G_{298}$(h–a)	$\Delta_r G_{298}$(f–a)
2	152.1	86.1	80.6	−158.1	29.9
3	136.8	82.3	53.7	−15.4	14.2
4	107.1	70.2	36.9	−65.2	36.9
5	140.7	87.7	43.0	−93.1	43.8

According to the Gibbs free energy values, the reaction of hydropyran cations formation is thermodynamically favorable.

The calculated values of the Gibbs activation energy reactions show that *trans*-2-butene is the most difficult to enter this reaction, which corresponds to the published data [8]. In the considered range of alkenes the smallest value of the activation energy is characteristic of the isobutylene reaction.

The transition states (*2e–5e*) correspond to the transition states of a water molecules separation stage. π-cations (*2b–5b*) are isomerized through the transition states (*2e–5e*) into σ-cations (*2f–5f*), which represent alk-oxycarbenium ions stabilized with water. According to the calculated Gibbs energy values, σ-cations (*2f–5f*) are more stable than the reactants and π-cations (Figs. 10.7 and 10.8).

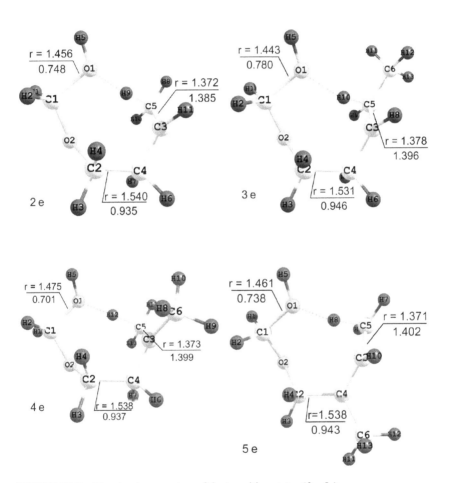

FIGURE 10.7   Structural parameters of the transition states (*2e–5e*).

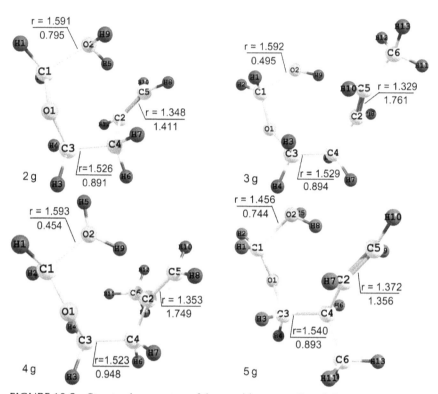

FIGURE 10.8   Structural parameters of the transition states (*2 g–5 g*).

The transition states (*2 g–5 g*) correspond to the process of the pyran cycle formation. The structures of the transition states (*2 g–5 g*) are shown in Fig. 10.8.

According to the calculated data, the formation of hydrogenated pyrans from the FD and alkenes is of an asynchronous nature.

10.4   CONCLUSION

Due to the analysis of the calculated data of the transition states structure of 4-alkyl-1,3-dioxane formation from formaldehyde oligomers and alkenes, it is found that 1,3-dioxane structures are formed in the result of direct isomerization of π-cation on the first stage. A free σ-cation formation is not observed here.

It is shown that interaction of formaldehyde oligomers with alkenes in gas phase or nonpolar solvents, accompanied by 1,3-dioxane formation must occur as a pseudo synchronous syn-addition. Only formaldehyde oligomers addition to ethylene can be considered as a synchronous interaction.

The highest activation energy is observed in case of the ethylene reaction. The introduction of alkyl substituents at the double bond leads to lower activation energy.

The calculations show that the formation of hydrogenated pyrans can be carried out by FD and alkenes reaction accompanied by 1,3-dioxanes formation as well. However, the activation energies of these reactions are higher than the ones of 1,3-dioxanes formation

KEYWORDS

- **1,3-Dioxanes**
- **AB Initio Calculation**
- **Alkenes**
- **Formaldehyde Oligomers**
- **Hydrogenated Pyrans**
- **Prins Reaction**
- **Transition State**

REFERENCES

1. Carballo, R. M., Ramirez, M. A., Rodriguez, M. L., Martin, V. S., & Padron, J. I. (2006). Iron(III)-promoted aza-prins-cyclization, Direct synthesis of six-membered azacycles. Org Lett 8, 3837–3840.
2. Bach, T., & Löbel, J. (2002) Selective Prins Reaction of Styrenes and Formaldehyde Catalyzed by 2,6-Di-tert-butylphenoxy(difluoro)borane. Synthesis 17, 2521–2526
3. Bogatyrev, V. F. (2008). Patent RU 2(330 010) C2. Way of Isoprene Obtaining. 21.
4. Chavre, S. N., Choo, H., Lee, J. K., Pae, A. N., Kim, Y., Cho, Y. S. (2008). 5- and 6-Exocyclic Products, cis-2,3,5-Trisubstituted Tetrahydrofurans, and cis-2,3,6-Trisubstituted Tetrahydropyrans via Prins-Type Cyclization. J. Org. Chem. 73, 7467–7471.
5. Yadav J. S., Reddy, B. V. S., Kumar, G. G. K. S. N., Aravind, S. (2008). The Aqueous Prins-Cyclization: A Diastereoselective Synthesis of 4-hydroxytetrahydropyran Derivatives. Synthesis 3, 395–400.

6. Aubele A. L., Lee, C. A., Floreancig, P. E. (2003). The "aqueous" Prins reaction. Org Lett 23, 4521–4523.
7. Tian, G.-Q., Shi, M. (2007). 2-(Arylmethylene)cyclopropylcarbinols could be converted to the corresponding tetrahydropyrans stereoselectively in the presence of Brønsted acids under mild conditions. A plausible Prins-type reaction mechanism has been proposed. Org. Lett. 9, 2405–2408.
8. Pastor, I. M., Yus, M. (2007). The Prins Reaction: Advances and Applications. Curr. Org. Chem. 10, 925–957.
9. Talipov, R. F. (1984). Dihydro- and alkeniltetrahydropyrans in the Prins Reaction. Ph.D. Thesis. Bashkir State University, Russia.
10. Vakulin, I. V. (2002). Research of heterocycle formation by the Prins reaction and their other transformations. Ph.D. Thesis. Bashkir State University, Russia.
11. Shepelevich, I. S., Talipov, R. F. (2003). Cyclization of cyclohexene with formaldehyde in trifluoroacetic acid, Bashkir Khim. J. 10, 58–60.
12. Olier, C., Kaafarani, M., Gastaldi, S., Bertrand, M. P. (2010). Synthesis of tetrahydropyrans and related heterocycles via prins cyclization; extension to aza-prins cyclization. Tetrahedron 66, 413–445.
13. Ibatullin, U. G., Talipov, R. F., Faizrakhmanov, I. S., Safarov, M.G. (1985). Ketones in an induced Prins reaction. Chem Heterocycl Compd 21:1392
14. Overman, L. E., Pennington, L. D. (2003). Strategic Use of Pinacol-Terminated Prins Cyclizations in Target-Oriented Total Synthesis. J. Org. Chem. 68, 7143–7157.
15. Taber, D. F. (2006). Stereoselective Construction of Oxygen Heterocycles. Org Chem Highlights.
16. Granovsky, A. A. http://classic.chem.msu/gran/gamess/index.html.
17. Hehre, W. J., Radom, L., Schleyer, P. R., Pople, J. A. (1986). Ab Initio Molecular Orbital Theory. Wiley, New York.
18. Freeman, F., Kasner, M. L., Hehre, W. J. (2001). An ab initio molecular orbital theory study of the conformational free energies of 2-methyl-, 3-methyl-, and 4-methyltetrahydro-2H-pyran, J. Mol. Struc. Theochem. 574, 19–26.
19. Kestutis, A., Mikkelsen, K. V., Stephan, P. A. (2008). Sauer On the Accuracy of Density Functional Theory to Predict Shifts in Nuclear Magnetic Resonance Shielding Constants due to Hydrogen Bonding. J. Chem. Theory Comput 4, 267–277.
20. Hehre, W. J., Radom, L., Schleyer, P. V. R., Pople, J. A. (1985). Ab Initio Molecular Orbital Theory. Wiley, New York.
21. (a) Muller, K. (1980). Reaction Paths on Multidimensional Energy Hypersurfaces. Angew Chem, Int Ed Engl 19:1–13; b) Schmidt, M. W., Gordon, M. S., Dupuis, M. (1985). The intrinsic reaction coordinate and the rotational barrier in silaethylene. J. Am. Chem. Soc. 107, 2585–2589. c) Garrett, B. C., Redmon, M. J., Steckler, R., Truhlar, D. G., Baldridge, K. K., Bartol, D., Schmidt, M. W., & Gordon, M. S. (1988). Algorithms and accuracy requirements for computing reaction paths by the method of steepest descent. J Phys Chem 92, 1476–1488. d) Baldridge, K. K., Gordon, M. S., Steckler, R., & Truhlar, D. G. (1989). Ab initio reaction paths and direct dynamics calculations. J., Phys Chem 93, 5107–5119.e).Gonzalez, C., Schlegel, H. B. (1989). An improved algorithm for reaction path following. J Chem Phys 90, 2154–2161.
22. Castillo, R., Moliner, V., & Andrés, J. (2000). A theoretical study on the molecular mechanism for the normal Reimer-Tiemann reaction. Chem. P Lett 318, 270–275.

23. (a) Giambagi, M., Giambagi, M., Grempel, D. R., & Heymann, C. D. (1975). Sur la définition d'un indice de liaison (TEV) pour des bases non orthogonales. Propriétés et applications. J. Chim Phys 72, 15–22.b) Mayer, I. (1983). Charge, bond order and valence in the AB initio SCF theory. Chem Phys Lett 97, 270–274.
24. Olah, G. A., & Schleyer, P. v. R. (1968). Carbonium Ions General aspects and methods of investigation Wiley-Interscience Publishers, New York.
25. Olier, C., Kaafarani, M., Gastaldi, S., & Bertrand, M. P. (2010). Synthesis of tetrahydropyrans and related heterocycles via prins cyclization, extension to aza-prins cyclization. Tetrahedron 66(2), 413–445.

CHAPTER 11

QUANTUM CHEMICAL INVESTIGATION OF FEATURES OF FORMATION OF O-CONTAINING HETEROCYCLES BY PRINS REACTION

I. V. VAKULIN, R. F. TALIPOV, O. YU. KUPOVA, G. R. TALIPOVA, A. V. ALLAGULOVA, and R. R. FAIZULLINA

CONTENTS

ABSTRACT

The role of formaldehyde dimer in O-containing heterocycles formation by the Prins reaction have been investigated. It was shown that the 1,3-dioxanes, hydrogenated pyrans and oxetanes can be can be obtained from formaldehyde dimers and alkenes in the gas phase. The activation energy of these reactions is different. It is lower for 1,3-dioxanes formation, and higher for oxetanes formation. Thus formation of 1,3-dioxanes happens in the conditions of kinetic control. Opposite, the hydrogenated pyrans formation happens in the conditions of thermodynamic control

11.1   INTRODUCTION

The Prins reaction [1] is a convenient technique of oxygen-containing heterocycles formation. It is applied in large-tonnage organic synthesis and total synthesis of complex natural products as well [2]. However, the formation of 4,4-dimethyl-1,3-dioxane along with fair quantities of by-products reduces practical value of this reaction. Thereupon, the clarification of the Prins reaction mechanism aimed at increasing selective ways of 1,3-dioxanes formation seems to be quite an urgent task.

Generally accepted mechanisms of products formation by Prins reaction demonstrated on Fig. 11.1.

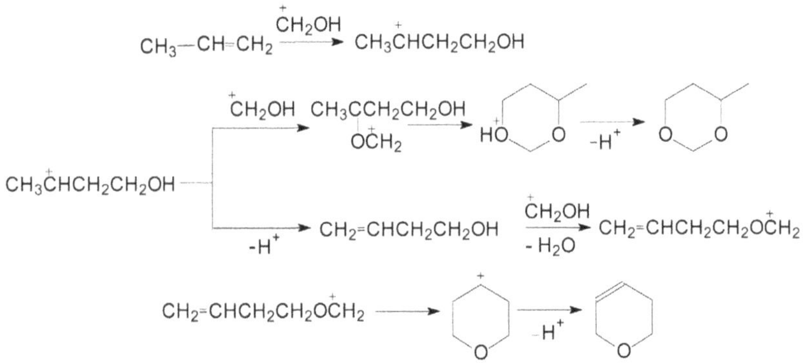

FIGURE 11.1   The Prins reaction mechanism [3].

Experimental [4] and theoretical data [5], [6] clearly demonstrate that the presence of formaldehyde oligomers (FO) is a prerequisite for the 1,3-dioxanes formation, whereas FO are much more reactive than the monomer. It is assumed that the FO addition at the double bond can be described as sequential, pseudo synchronous or concert [7]. The implementation of the two mechanisms corresponds to the experimental data on a stereoselective Prins reaction for a series of cycloalkenes [8].

However, the exact nature of the addition and its relationship to the alkenes structure are still poorly described.

Therefore, the aim of our study was to clarify the mechanism of alkenes and FO interaction using quantum chemical methods. Thereto we determined thermodynamic parameters of the reactions ($\Delta_r H^0_{298}$ and $\Delta_r G^0_{298}$), found the key intermediates and transition states of the corresponding transformations and considered characteristics of their structure depending on the alkene structure involved in the reaction. The following alkenes were used as model compounds: ethylene, propylene, butene-1, isobutylene and t-butene-2. A formaldehyde dimer (FD) served an example for FO calculation.

11.2 METHODS OF CALCULATIONS

MOPAC-97 and PC GAMESS v7.1 programs [9] were used in carrying out quantum and chemical calculations.

Semiempirical calculations were carried out in the MNDO approximation in the AM1 parameterization [10]. Ab initio calculations were carried out using a split-valence basis set 6–31G with the d-polarization function for all atoms [11]. The electron correlation was considered by using the Møller-Plesset second order perturbation theory with a frozen skeleton of electrons (frozen core, FC) [13].

Searching for equilibrium geometry of the transition states was carried out in the MOPAC program using the SADDLE procedure in AM1 approximation with subsequent refinement of geometry using the TS procedure.

The initial geometry of the transition states was optimized in the GAMESS program with approximation of MP2(fc)/6–31G(d) by setting RUNTYP = SADPOINT under $CONTRL.

Verification of the transition state was made by calculating vibrational parameters for the obtained transition state geometry with the subsequent analysis of the calculated frequencies of the IR spectrum, as well as modeling the found transition state into the starting materials and final products by means of the IRC procedure.

11.3 CALCULATION RESULTS AND DISCUSSION

Earlier in studying structural features of intermediates, prereaction complexes and transition states we showed [12] that the interaction of alkenes and FO on the first stage leads to the formation of several isomeric π-cations with one capable of further transformations only. The analysis of the potential energy surface section corresponding to the reaction coordinate shows that FO adding with 1,3-dioxanes formation can be regarded as a pseudo synchronous interaction only in case of ethylene. For other alkenes it more corresponds to the sequential double bond FO addition. However, it is impossible to reveal the existence of a stable σ-cation in transforming a reactive π-cation into 1,3-dioxane. In this case, according to calculations, the alkene and FO interaction in a gas phase or nonpolar media must proceed as a syn-addition.

In this chapter, we have determined the thermodynamic parameters of the reactions corresponding to different variants of FD and alkenes interaction which is accompanied by oxygen-containing heterocycles formation. Their comparative analysis is conducted either.

11.3.1   1,3-DIOXANES

In Fig. 11.2, the potential energy surface section corresponding to 1,3-dioxanes formation from FD and alkenes is shown:

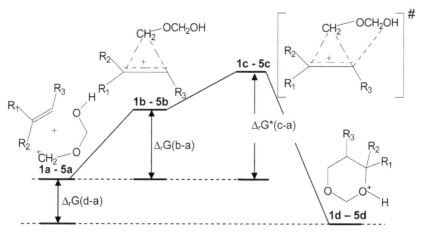

FIGURE 11.2 Energy profile of alkenes and formaldehyde dimer reaction, where R1 = R2 = R3 = H (1); R1 = CH3, R2 = R3 = H (2); R1 = C2H5, R2 = R3 = H (3); R1 = R2 = CH3, R3 = H (4); R1 = H, R2 = R3 = CH3 (5).

The calculated values of the corresponding thermodynamic parameters of the reactions are represented in Table 11.1.

TABLE 11.1 Energy Parameters of the Key Stages of 1,3-Dioxanes Formation From Alkenes and FD, kJ/mol

Alkene	$\Delta_r G^{\neq}_{298}$	$\Delta_r G_{298}(b\text{–}a)$	$\Delta_r G_{298}(d\text{–}a)$	$\Delta_r G_{298}(c\text{–}b)$
1	138.7	110.9	−7.4	27.7
2	99.2	80.6	−27.6	6.6
3	72.2	53.7	−30.5	18.5
4	66.2	33.2	−33.8	32.9
5	114.7	43.0	−35.3	71.6

According to the Gibbs free energy values, a dioxane (1d–5d) formation is thermodynamically favorable; it slightly increases while the number of alkyl substituents at the double bond grows. Thus, the Gibbs free energy value for ethylene amounts to −7.4 kJ/mol, whereas this value for butene-2 is −35.3 kJ/mol.

The Gibbs activation energy for ethylene obtains the highest value (138.7 kJ/mol), and indeed there are no examples of ethylene involvement into the Prins reaction.

Introduction of alkyl substituents at one end of the double bond leads to a significant decrease in activation energy (up to 99.2 kJ/mol). Introducing alkyl substituents at both ends of the double bond lowers the activation energy that can be explained by a steric factor.

11.3.2   HYDROGENATED PYRANS

According to literature [5], hydrogenated pyrans are thermodynamically more favorable reaction products as compared with 1.3-dioxans (Fig. 11.3). At the same time it is assumed that their formation occurs with the participation of formaldehyde monomer. However, the results of our calculations show that hydrogenated pyrans can also be produced from FO and alkenes.

Together with alkenes (2a–5a) FD preliminarily forms the corresponding π-cations (2b–5b), which are isomerized through the transition states (2e–5e) into σ-cations (2f–5f). σ-cations (2f–5f) are subsequently transformed via the transition states (2 g–5 g) into the pyran cycle (2h–5h). The calculated values of thermodynamic parameters of the corresponding reaction stages are presented in Table 11.2.

TABLE 11.2   Energy Parameters of the Hydrogenated Pyrans Formation Reactions, kJ/mol

Alkene	$\Delta_r G^{\neq}{}_{298}$	$\Delta_r G^{\neq}{}_{298}(e \rightarrow a)$	$\Delta_r G_{298}(b \rightarrow a)$	$\Delta_r G_{298}(h \rightarrow a)$	$\Delta_r G_{298}(f \rightarrow a)$
2	152.1	86.1	80.6	−158.1	29.9
3	136.8	82.3	53.7	−15.4	14.2
4	107.1	70.2	36.9	−65.2	36.9
5	140.7	87.7	43.0	−93.1	43.8

According to the Gibbs free energy values, the reaction of hydropyran cations formation is thermodynamically favorable.

The calculated values of the Gibbs activation energy reactions show that t-butene-2 is the most difficult to enter this reaction, which corresponds to the published data [8]. In the considered range of alkenes the smallest value of the activation energy is characteristic of the isobutylene reaction.

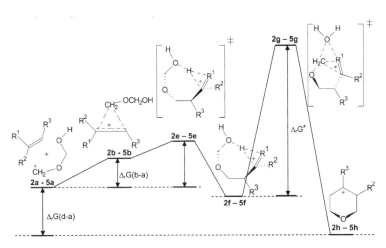

FIGURE 11.3   The scheme of hydrogenated pyrans formation from alkenes and FD.

The transition states (2e–5e) correspond to the transition states of a water molecules separation stage. π-cations (2b–5b) are isomerized through the transition states (2e–5e) into σ-cations (2f–5f), which represent alkoxycarbenium ions stabilized with water. According to the calculated Gibbs energy values, σ-cations (2f–5f) are more stable than the reactants and π-cations (Fig. 11.4).

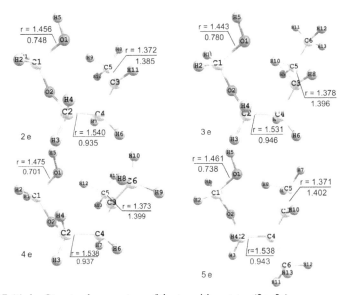

FIGURE 11.4   Structural parameters of the transition states (2e–5e).

The transition states (2 g–5 g) correspond to the process of the pyran cycle formation. The structures of the transition states (2 g–5 g) are shown in Fig. 11.5.

According to the calculated data, the formation of hydrogenated pyrans from the FD and alkenes is of an asynchronous nature.

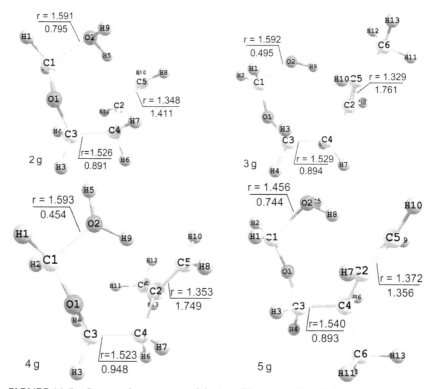

FIGURE 11.5 Structural parameters of the transition states (2 g–5 g).

11.3.3 OXETANES

Unpredictably due to the results of our calculations, the formation of oxetanes can also be accomplished with formaldehyde olygomers participation and 1,3-dioxanes and hydrogenated pyrans formation (Fig. 11.6).

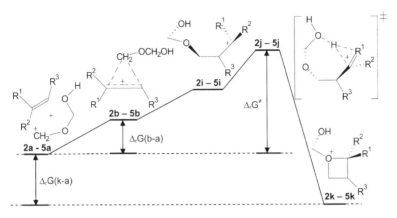

FIGURE 11.6   Formation of oxetanes from alkenes and formaldehyde dimers.

According to the scheme, formaldehyde dimers preliminary form the corresponding π-cations (2b–5b) with alkenes (2a–5a), which are isomerized into σ-cations (2i–5i). These σ-cations are transformed through the transition states (2j–5j) into oxetanes (2k–5k). The calculated values of the thermodynamic parameters are represented in Table 11.3.

TABLE 11.3   Energy Parameters of Oxetane Formation Reaction, kJ/mol

Alkenes	$\Delta_r G^{\neq}_{298}$	$\Delta_r G_{298}(b \rightarrow a)$	$\Delta_r G_{298}(k \rightarrow a)$	$\Delta_r G_{298}(j \rightarrow b)$
2	270.5	92.0	15.7	178.2
3	197.5	53.7	5.2	143.8
4	98.2	36.9	4.5	64.9
5	188.5	43.0	5.3	143.6

The calculated values of the Gibbs activation energy reaction show that propylene is the worst reactive here. The activation energy possesses the lowest value when the substituted alkenes (isobutylene and t-butene-2) are involved into the reaction.

The transition states of the oxetane formation reaction are shown in Fig. 11.7.

Thus, according to the calculated data, hydrogenated pyrans and oxetanes are possible to be formed along with the formation of alkyl-substituted 1,3-dioxanes in the reaction of FD with alkenes. In this case the formation of 1,3-dioxanes is characterized by lower activation energy

as compared with the energy of hydrogenated pyrans and oxetanes formation reactions.

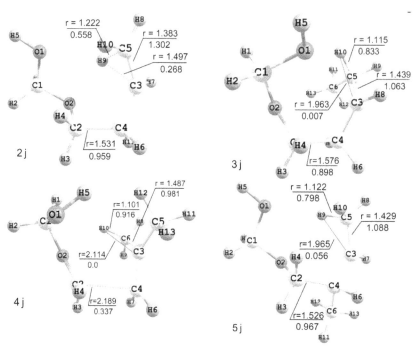

FIGURE 11.7   Structural parameters of the transition states of the oxetane formation
reaction.

11.4   CONCLUSION

The quantum chemical modeling revealed that the formation of 1,3-di-
oxanes from FO and alkenes in the gas phase or nonpolar media is represented as a pseudo synchronous syn-addition.

The highest activation energy is observed in case of the ethylene reaction. The introduction of alkyl substituents at the double bond leads to
lower activation energy.

The calculations show that the formation of hydrogenated pyrans and
oxetans can be carried out by FD and alkenes reaction accompanied by

1,3-dioxanes formation as well. However, the activation energies of these reactions are higher than the ones of 1,3-dioxanes formation.

KEYWORDS

- **AB Initio Calculation**
- **Alkylated 1,3-Dioxanes**
- **Formaldehyde Oligomers**
- **Hydrogenated Pyrans**
- **The Prins Reaction**
- **Transition State**

REFERENCES

1. Aubele, A. L., Lee, C. A., & Floreancig, P. E. (2003). The "Aqueous" Prins Reaction. Org. Lett. 5(23), 4521–4523.
2. Isidro, M., Pastor, & Miguel Yus. (2007). The Prins Reaction, Advances and Applications Current Organic Chemistry 11(10), 925–957(33).
3. Arundale, E., & Mikeska, L. A. (1952). The Olefin-Aldehyde Condensation. The Prins Reaction Chem. Rev. 51(3), 505–555.
4. Talipov, R. F., & Safarov, M. G. (1997). Bashkir. Khim. Zh. 4(3), 10–16.
5. Vakulin, I. V. C. (2002). Sci. (Chem.) Dissertation. Moscow.
6. Shepelevich, I. S., & Talipov, R. F (2003). Bashkir. Khim. Zh. 10(4), 58–60.
7. Overman, L. E., & Pennington, L. D. (2003). Strategic Use of Pinacol-Terminated Prins Cyclizations in Target-Oriented Total Synthesis. J. Org. Chem. 68(19), 7143–7157.
8. Carballo, R. M., Ramirez, M. A., Rodriguez, M. L., Martin, V. S., & Padron, J. I. (2006). Iron (III)-promoted aza-Prins-cyclization, Direct Synthesis of Six-Membered Azacycles, Org. Lett. 8, 3837–3840.
9. Granovsky, A. A., http://classic.chem.msu. su/gran/gamess/index.html.
10. Dewar, M., & Thiel, W. (1977). J. Amer. Chem. Soc. 99:4499.
11. Hehre, W. J., Radom, L., Schleyer, P. V. R., & Pople, J. A. (1986). Ab Initio Molecular Orbital Theory, Wiley, New York.
12. Vakulin, I. V., Kupova, O. Yu, & Talipov, R. F. (2010) Vestn. Bashkir. Univer. 15(2), 294–297.
13. Kestutis Aidas, Kurt, V., & Mikkelsen Stephan, P. A. (2008). Sauer On the Accuracy of Density Functional Theory to Predict Shifts in Nuclear Magnetic Resonance Shielding Constants due to Hydrogen Bonding J., Chem. Theory Comput., 4(2), 267–277.

PART III

NEW TECHNOLOGICAL SOLUTIONS IN THE PRODUCTION OF POLYMERS

CHAPTER 12

MODIFICATION OF A NEODYMIUM CATALYST FOR SYNTHESIS POLYISOPRENE

V. P. ZAKHAROV, V. Z. MINGALEEV, E. M. ZAKHAROVA,
I. D. ZAKIROVA, and L. A. MURZINA

CONTENTS

ABSTRACT

A procedure was suggested for intensifying the complexation of neodymium chloride with isopropyl alcohol in preparation of the neodymium catalyst for polyisoprene synthesis by using a small-size tubular turbulent diffuser-confuser apparatus in the step of the reaction mixture formation.

12.1   INTRODUCTION

Intensification of the complexation of neodymium chloride with isopropyl alcohol by preliminary mixing of the reaction mixture in a tubular turbulent diffuser-confuser apparatus was studied. The procedure ensures acceleration of the process, incorporation of a larger amount of isopropyl alcohol in the complex, and high activity of the neodymium catalyst in the polyisoprene synthesis.

Catalytic systems based on lanthanide compounds are effective catalysts of polymerization of dienes (mainly of isoprene and butadiene). A distinctive feature of these systems is high cis stereospecificity. In particular, the synthesized polybutadiene and polyisoprene contain approximately 98% 1,4-cis units [1].

Catalytic systems formed by the reaction of lanthanide chloride (catalyst) with an organoaluminum compound (cocatalyst) exhibit low activity in diene polymerization [1, 2]. The catalyst activity is considerably enhanced on introducing electron-donor compounds [2]. An electron-donor ligand increases the extent of the covalence of the lanthanide-halogen bond and thus favors alkylation with an organoaluminum compound with the formation of lanthanide-carbon bonds active in polymerization [1, 2]. Thus, lanthanide catalytic systems highly active in diene polymerization are formed in the reaction of an organoaluminum compound with neodymium chloride in the form of a complex compound with an electron-donor ligand.

As shown in Ref. [3], the complex of neodymium chloride with tributyl phosphate (TBP) is formed slowly. The process is accompanied by essential changes in the structure of the crystal lattice of both the initial $NdCl_3$ and the complex $NdCl_3 nTBP$. The complexation with the organic ligand results in loosening of the crystal structure of neodymium chloride and in formation of an amorphous precipitate of $NdCl_3 nTBP$ [4]

Today isoprene rubber is commercially produced in Russia using alcohol complexes of neodymium chloride [5]. In this connection, we examined in this study the possibility of intensifying the complexation of neodymium chloride with isopropyl alcohol by preliminarily mixing the initial reaction mixture in a turbulent diffuser-confuser apparatus to obtain a highly active catalyst of stereospecific polymerization of isoprene. The results obtained are prerequisites for improvement of the process for preparing gel-free neodymium isoprene rubber.

The neodymium component of the catalytic complex was prepared from the hydrate $NdCl_3$-$0.6H_2O$ (hereinafter, $NdCl_3$) with isopropyl alcohol (IPA) in liquid paraffin at 25°C. The reactant molar ratio was $NdCl_3$: IPA =1 : 3. The reaction of $NdCl_3$ with IPA was performed by two methods. In method 1, the complexation occurred in a large-volume flask with slow stirring throughout the process. In method 2, the starting reactants were preliminarily mixed in the step of the reaction mixture formation in a tubular turbulent diffuser-confuser reactor [6] or 2–3 s, after which the reaction mixture was fed to the flask in which the conditions were the same as in method 1. The design of the six-section tubular turbulent apparatus [diameter of the wide part (diffuser) 0.024 m, diameter of the narrow part (confuser) 0.015 m, length of the diffuser-confuser section 0.048 m, diffuser opening angle 45°] corresponded to the maximal dissipation of the specific kinetic energy of the turbulence, taking into account the process specificity.

The dynamics of the isopropyl alcohol consumption (content in the liquid phase) in the course of complexation was monitored by high-performance liquid chromatography (Shimadzu LC-20 chromatograph). To prepare the catalytic complex of the composition $NdCl_3$: triisobutylaluminum (TIBA) . piperylene – 1 . 12 . 1 (molar ratio), a hermetic 500 cm³ reactor was charged in a dry argon stream with calculated amounts of toluene (solvent), piperylene, and TIBA. Then the mixture was cooled under stirring to −15°C, after which the alcohol complex of neodymium chloride was added with stirring over a period of 40 min at −5°C. After that, the mixture was kept for 24 h at 25°C with intermittent stirring to prepare the catalytic complex. The isoprene polymerization was performed in toluene in hermetically sealed ampules under argon at 25°C. The polymer yield was monitored gravimetrically. The particle size of the initial neodymium chloride, its alcohol complex, and the catalytic system was examined by laser diffraction or scattering with a Sald-7101 device (Shimadzu).

In the initial reaction mixture, the NdCl$_3$ particle size is about 15 pm (Fig. 12.1). In the first 4 h of stirring, when performing the complexation by method 1, the neodymium chloride particle size decreases to 4 pm. Further increase in the process time does not noticeably affect the particle size of the initial neodymium chloride.

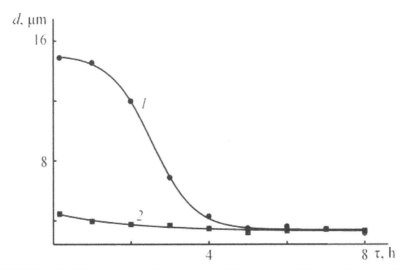

FIGURE 12.1 Mean diameter d of particles of the starting NdCl$_3$ as a function of the time τ of complexation with isopropyl alcohol. Method: (1) 1 and (2) 2; the same for Figs. 12.2–12.4.

Single circulation of the starting reactants in the step of formation of the reaction mixture (method 2) leads to disintegration of neodymium chloride particles to the size characteristic of the suspension formed in 4–5 h of the synthesis by method 1 (Fig. 12.1). With the complexation performed for more than 5 h, the neodymium chloride particle size is about 4 pm irrespective of the synthesis method. Thus, short hydrodynamic action in the turbulent mode on a suspension containing the starting NdCl$_3$ considerably intensifies the dispersion of the solid phase particles in complexation with the alcohol.

In the course of the complexation, we took samples of the reaction mixture in which we observed clear segregation of the system into three layers: I (lower layer), a violet suspension of the starting NdCl$_3$; II (middle layer), the complex NdCl$_3$·nIPA in the form of a white paste; and III (upper layer), colorless liquid paraffin.

Therefore, the kinetics of the accumulation of the complexation product can be estimated by quantitative calculation of the volume fraction ϕ of layer II in the reaction mixture, which correlates with the $NdCl_3 \cdot nIPA$ yield.

Because of the topochemical nature of the process, its rate is determined by the particle size of the starting $NdCl_3$. The kinetic curve of the $NdCl_3 \cdot nIPA$ accumulation in the synthesis by method 1 is S-shaped (Fig. 12.2). Analysis of the dynamics of the $NdCl_3$ particle size variation and of the $NdCl_3 \cdot nIPA$ accumulation shows that the complexation acceleration is observed on reaching a definite particle size of the starting neodymium chloride. Short hydrodynamic action on the reaction system at the moment of its formation considerably increases the yield of the complex owing to the dispersion of the $NdCl_3$ particles.

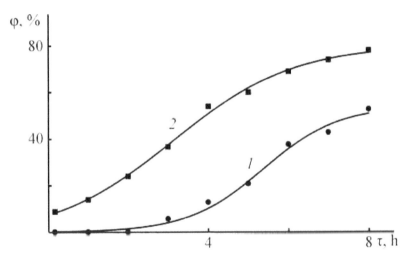

FIGURE 12.2   Kinetic curves of accumulation of the complex $NdCl_3 \cdot nIPA$ in the reaction mixture. (j) yield of $NdCl_3$-n IPA and (**t**) synthesis time.

Apparently, a significant decrease in the neodymium chloride particle size in the first 4–5 h of the synthesis is not due to the occurrence of the complexation on the particle surface (contracting sphere model), because the yield of the complex in this period is low. Most probably, a decrease in the particle size of the initial $NdCl_3$ is associated with disintegration of the solid phase under the action of mechanical stirring, with the wedging effect produced by solvent molecules solvating the particle surface. This is confirmed by a considerable decrease in the $NdCl_3$ particle size and by accelerated accumulation in the reaction mixture of the complex $NdCl_3IPA$ at short hydrodynamic action on the reaction mixture in the initial moment of the complexation.

Performing the process by method 1 favors formation of $NdCl_3 \times IPA$ particles of size in the range 50–160 nm (Fig. 12.3). The use of a tubular turbulent apparatus allows synthesis of the complex with considerably narrower particle size distribution (interval 50–65 nm). A decrease in the $NdCl_3 \times IPA$ particle size both in method 1 and in method 2 favors incorporation of larger amounts of isopropyl alcohol into the complex. This is apparently due to an increase in the specific surface area of the solid phase particles, ensuring the access of isopropyl alcohol for the solvation. With preliminary mixing of the reaction mixture in a tubular turbulent apparatus, larger amount of isopropyl alcohol is incorporated into the complex. In particular, it becomes possible to synthesize the complex $NdCl_3 2,5IPA$, which cannot be formed in the synthesis by method 1, despite the fact that, with this procedure, it is possible to disperse particles to a mean size of the order of 50 nm.

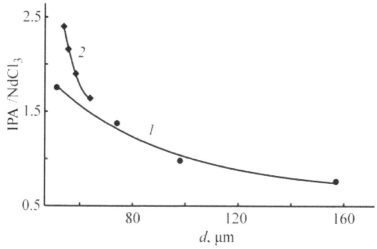

FIGURE 12.3 Composition of the complex $NdCl_3 \cdot nIPA$ as a function of the diameter of its particles d.

The complexes $NdCl_3 \times IPA$ synthesized by different methods were used for preparing the catalytic system $NdCl_3 \cdot nIPA$-TIBA-piperylene, active in stereospecific polymerization of isoprene. A specific feature of the formation of this catalyst is coarsening of the dispersed phase particles compared to the complex $NdCl_3{}^\wedge IPA$. In particular, the particle size of the alcohol complexes of neodymium chloride prepared by method 1 in 2, 5, and 8 h is 160, 80, and 50 nm, respectively. At the same time, the catalytic system $NdCl_3 \times IPA$-TIBA-piperylene prepared on the basis of these alcohol complexes is characterized by the mean particle size of 0.15 pm. The particle size of the complex $NdCl_3 \cdot nIPA$ synthesized by method 2 in 2, 5, and 8 h is 75, 53, and 51 nm, respectively, whereas in the catalytic complex the particle size is as large as 0.12 pm.

The dependence of the initial rate of the isoprene polymerization in the presence of the neodymium catalyst on the composition of the complex $NdCl_3 \cdot nIPA$ is S-shaped (Fig. 12.4). The dependence is common for the neodymium component prepared by different methods 1 and 2. The catalyst activity considerably increases with an increase in the IPA content from 1.4 to 2 mol per mole of $NdCl_3$. The complexation of $NdCl_3$ with isopropyl alcohol at slow stirring (method 1) leads to the formation of $NdCl_3$ 1,65IPA in 8 h. Because longer complexation does not leads to a further increase in the degree of solvation with isopropyl alcohol, this neodymium catalyst is characterized by the limiting, under the given experimental conditions, initial rate of the isoprene polymerization, $W_0 = 0.017$ L mol^{-1} min^{-1}. The isoprene polymerization rate can be additionally increased by using a tubular turbulent apparatus in the step of formation of the reaction mixture in solvation of neodymium chloride with isopropyl alcohol (method 2). At the maximal degree of IPA incorporation (2.55 mol of IPA per mole of $NdCl_3$), the initial rate of the isoprene polymerization is $W_0 = 037$ L mol^{-1} min^{-1}, which exceeds by a factor of 2.2 the similar parameter attained with $NdCl_3 \cdot nIPA$ synthesized by method 1. Figures 12.3 and 12.4 show that the enhancement of the catalyst activity is mainly associated with the degree of the $NdCl_3$ solvation with isopropyl alcohol. At the same time, we should not rule out the possibility that the more active catalyst is formed owing to a decrease in the particle size of the $NdCl_3 \cdot nIPA$-TIBA-piperylene complex from 0.15 to 0.12 pm in going from method 1 to method 2.

Thus, we revealed a nontraditional way of intensifying the complexation of neodymium chloride with isopropyl alcohol by single circulation of the reaction mixture through a tubular turbulent diffuser-confuser apparatus. The suggested synthesis procedure with short hydrodynamic action in the step of formation of the reaction mixture leads to considerable acceleration of the complexation, to incorporation of a larger amount of isopropyl alcohol into the complex, and to the subsequent formation of the more finely dispersed catalytic complex $NdCl_3 \cdot nIPA$-TIBA-piperylene, which is manifested in high activity of the neodymium catalyst in polyisoprene synthesis.

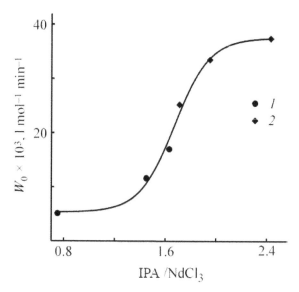

FIGURE 12.4   Initial rate W_0 of isoprene polymerization as a function of the composition of the complex $NdCl_3 nIPA$ used for preparing the lanthanide catalyst.

12.2   CONCLUSION

A procedure was suggested for intensifying the complexation of neodymium chloride with isopropyl alcohol in preparation of the neodymium catalyst for polyisoprene synthesis by using a small-size tubular turbulent diffuser-confuser apparatus in the step of the reaction mixture formation.

The activity of the neodymium catalyst is mainly determined by the degree of isopropyl alcohol incorporation in the complex $NdCl_3 nIPA$. Under the experimental conditions, the highest activity in isoprene polymerization is exhibited by the complex $NdCl_3$-(2.0–2.5)IPA, which can be prepared by single circulation of the reaction mixture through the turbulent apparatus.

12.3   ACKNOWLEDGMENTS

The study was financially supported by the Russian Foundation for Basic Research (project no. 11-03-97017).

KEYWORDS

- **Active Sites**
- **Isoprene**
- **Molecular Characteristics**
- **Neodymium Catalysts**
- **Polymerization**
- **Tubular Turbulent Pre-Reactor**

REFERENCES

1. Monakov, Yu. B., & Tolstikov, G. A., (1990). Kataliticheskaya polimerizatsiya 1,3-dienov (Catalytic Polymerization of 1,3-Dienes), Moscow Nauka.
2. Marina, N. G., Monakov, Yu. B., Sabirov, Z. M., & Tolstikov, G. A. (1991) Vysokomol. Soedin, Ser. A, 33(3), 467–475.
3. Korovin, S. S., Galaktionova, O. V., Lebedeva, E. N., & Voronskaya, G. N. (1975). Zh. Neorg. Khim, 20, 908–913.
4. Gallazi, M. C., Bianchi, F., Depero, L., & Zocc, M., (1988). Poly-mer, (8), 1516–1521.
5. Rakhimov, R., Kh., Kutuzov, P. I., Bazhenov, Yu. P., & Nasy-rov, I. (1997). Sh. Bashk. Khim. Zh, 4(2), 14–17.
6. Zakharov, V. P., Berlin, A. A., Monakov, Yu. B., & De- berdeev, R.Ya. (2008). Fiziko-khimicheskie osnovy protekaniya bys-trykh zhidkofaznykh protsessov (Physicochemical Principles of the Progress of Fast Liquid-Phase Processes), Moscow Nauka.

CHAPTER 13

OIL NEUTRALIZATION IN TURBULENT MICROREACTOR

F. B. SHEVLYAKOV, T. G. UMERGALIN, V. P. ZAKHAROV, I. D. ZAKIROVA, and E. M. ZAKHAROVA

CONTENTS

ABSTRACT

The regularities of crude oil preparation, particularly its neutralization by an alkali, is been considered in current study. In order to neutralize the crude oil at electric desaltation stage, it is proposed to use the high-performance compact tubular turbulent apparatus of confusor-diffuser design

13.1   INTRODUCTION

Hydrolysis of inorganic salts, which runs on electric desalting plant (EDP) is accompanied by the formation of an acidic environment. It is used to conduct the neutralization of such crude oil by alkalis as well as by organic amines. Because of the difference of the density and viscosity of the oil and alkaline solutions, the neutralization occurs not equimolarly under diffusion control and is accompanied by increase consumption rates of the neutralizing agent. The excess amount of the lye supplied to neutralize is reflected on the thermal oil-processing, reducing the activity of cracking catalysts. Reducing the impact of excess alkali can be achieved by intensification of mixing alkali point of application to the flow of oil.

It is known way of oil neutralization by aqueous alkaline solution, which for the dispersion of aqueous alkali proposes, in particular, carry out their partial premixing with oil (1%) [1], or use a different nozzle design [2]. Experience has shown that the effectiveness of both methods is low. In addition, they are characterized by disadvantages associated with the need to service the mixers and control when the flow of oil and/or the lye is changeable.

Another known on the electric desalting plants method [3] uses mixing valves, which process is working at high pressures drops across the valve, which is associated with significant energy costs to provide the required performance EDP on oil, moreover, the hydraulic losses will increase significantly during the works with heavy and viscous oils.

A more effective way is to neutralize the oil on electric desalting installation with prior addition of de-emulsifier, which produces a mixture of oil with a soda-alkaline solution in static mixer type like Sulzer SMV Hemiteh [4, 5]. Disadvantage of this method is the complexity of device design, large quantity of metal usage, high pressures drops of at high flows of oil.

The purpose of this study was developing a compact mixing reactor design, which can reduce the working pressure drop during the neutralization and amount of the alkaline agent required.

In order to solve the problem indicated, the oil neutralization was carried out in a turbulent reactor confusor-diffuser design [6]. In this case, the oil prepared and preheated to 110–120°C enters the inlet of the first section of a turbulent tubular reactor confusor-diffuser design with a flow rate of 680–750 m^3/h, where the dispersion of the two-phase system is occurring. Later, alkaline solution 1–2% by wt. is coaxially inserted in the first section of the reactor-mixer through the frontal atomizers. This allows improve the efficiency of oil neutralization due to a significant reduction of fresh and spent volumes of lye. The advantages of the device are absence of mixing devices, low-pressure drop, and low quantity of metal usage.

Due to possibility to increase the quality of alkali and oil mixing via fine dispersion and uniform distribution of the alkali in the entire volume of oil, the conditions for the creation of a homogeneous emulsion-phase model of system "liquid-liquid" in tubular turbulent apparatus were studied [7, 8].

Droplets distribution of the dispersed phase in size to the formation of fine homogeneous systems in the confusor-diffuser channels is narrowed by increasing speed of immiscible fluid streams. Increase in volumetric flow velocity ω and the number of diffuser confused N_c sections 1 to 4 leads to reduction of the volume-surface diameter of droplets of the dispersed phase and, consequently, to increase in the specific surface of the interface, which in the case of fast chemical reactions intensify the total process. Inadvisability of using the apparatus with the number of diffuser sections N_c confused over 5 ± 1, making these devices simple and inexpensive to manufacture and operate as well as compact, for example, length does not exceed 8–10 caliber (L/d$_D$).

There is a range of volume velocity of two-phase flow, which corresponds to the cone-channel confused with optimal diameter of the diffuser to confuser (further indicated as d_D/d_C). The distance is limited from bottom by seating stratified two-phase flow, and is limited from top by energy costs arising from the increased pressure on the ends of the device (Dp~w^2). In particular, the ratio $d_D/d_C = 3$ corresponds to the interval 44 < w < 80 cm^3/s, and $d_D/d_C = 1.6$ corresponds to the interval 80 < w < 180 cm^3/s, and further increase in the velocity of the dispersed system (w > 180 cm^3/s) determines the need to further reduce the ratio d_D/d_C until $d_D/d_C = 1$, that is, small units cylindrical structure are effective enough in this case. Thus, the flow, in which the dispersed particles are uniformly dispersed in the unit of confusor-diffuser design in comparison with the cylindrical channel, is formed at the lower velocities of the dispersed system, and the higher the ratio d_D/d_C, the lower the required value W (due to changing the value of the Reynolds number Re according to the ratio Re ~ d_D/d_C).

Thus, the change in the rate of fluid flows in tube W devices and relation d_D/d_C is almost the only, but very effective way to affect the nature of the dispersion and the quality of the emulsion. These patterns of relationships allow under optimal conditions and without nontechnical or technical problems create thin homogeneous dispersion systems "liquid-liquid" with a minimum residence time of the reactants in the mixing zone, and use simple apparatus to design small confusor-diffuser design.

Another important quantity that characterizes the quality of the emulsion is the polydispersity coefficient k. Ratio L_c/d_D almost no effects on the polydispersity of emulsions obtained. The increase in the spread of the dispersed phase in size is observed during increasing of the ratio d_D/d_C, and quite homogeneous emulsion is formed in the diffuser confused channels tubular device with $d_D/d_C = 1.6$. In particular, the value of k in $d_D/d_C = 1.6$ for $L_c/d_D = 2$–3 equals 0.72–0.75, whereas the k is reduced to 0.63 and 0.41 when the ratio d_D/d_C is 2 and 3, respectively.

Creation of intensive longitudinal mixing in a two-phase system in tubular turbulent apparatus with the ability to increase the surface of contact between the phases allows intensify the flow of fast chemical reactions at the interface.

The dependences obtained allow predict the dispersion of droplets of alkali in oil, which makes it possible to design a mixer for use in a wide range of flow rates of mixed liquids.

The process at low differential pressure is necessary to carry out for effective mixing of oil-base, which is directly related to the energy needed to provide the performance ELOU required. However, the hydraulic losses increase significantly when working with heavy and viscous oil.

The pressure drop is expressed by the relationship $\Delta P=(lL/d+x)\rho w^2/2)$, where ζ is a coefficient of local resistance, l is a friction coefficient, L is a length, d is diameter, r is a density, w is a speed.

The coefficient of local resistance to the unit area with sudden expansion is calculated (in the calculation of the velocity head speed in a smaller cross section) by the formula $(1-S_1/S_2)^2$ (Fig. 13.1), and cylindrical portion of the apparatus for $\zeta = 1$, while coefficient of local resistance to the plot device of a sudden contraction (in the calculation of the velocity head speed in a smaller section) $\zeta = 0.38$.

S_1

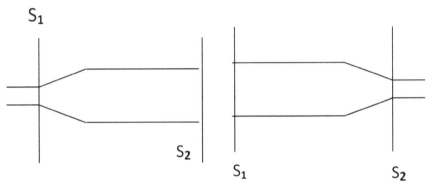

S_2

S_1

S_2

FIGURE 13.1 Scheme for calculation of coefficient of local resistance.

The values of friction coefficient for turbulent flow can be calculated by the Blasius formula: $l = 0.316/Re^{0.25}$.

The pressure drop in the section is the sum of the pressure drop in a smooth tube, expansion (diffuser) and narrowing (confuser) (Fig. 13.2).

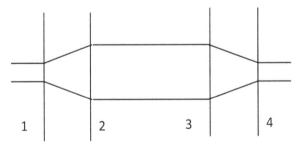

1   2   3   4

FIGURE 13.2 Scheme for calculation of pressure drop in tube.

$$\Delta P = (P_1 - P_2) + (P_2 - P_3) + (P_3 - P_4)$$

The total pressure drop is the sum of pressure drops in each section.

Calculation by these formulas was done according to experimental data of measuring the pressure at the ends of tubular turbulent apparatus consisting of 20 sections with a water flow. Comparison of calculated data obtained with respect to the model system shows correlations with the experimental data for the pressure in the apparatus: D P $_{practical}$ = 0.955 atm, D P $_{theoretical}$ = 1.062 atm. Calculation of diameter of the narrow section (confuser) section on the proposed formulas, based on the requirements for the pressure drop in the apparatus D P $\leq$ 0.6 atm, was done (Table 13.1).

TABLE 13.1 Calculation of the Diameter of the Diffuser at DC for Oil in the Apparatus of Confusor-Diffuser Design

$DP_{5\,section}$	0.6	atm
$DP_{section}$	0.118	atm
DP_{3-4}	6643	kgF/m²
DP_{2-3}	51.1	KgF/m²
DP_{1-2}	7875	KgF/m²
z_C	0.38	
z	0	
z_D	0.46	
r	762	Kg/m³
L	0.875	m
l_C	0.0087	
l_D	0.0115	
d_D	0.35	m
Re	177×10⁴	
Re_D	577×10³	
w_C	6.62	m/s
W	2.163	m/s
S_D	0.096	m²
S_C	0.031	m²
d_C	0.198	m

The pressure drop at the ends of the device with a diameter of confuser $d_C = 0.2$ m is $DP_{5unit} \approx 0.52$ atm, which is optimal for steel neutralization of oil.

Sharp rise in temperature is observing while using of concentrated solutions of the reagents during the neutralization of acidic environments. In this case, the small tubular turbulent reactors confusor diffuser designs define the ability to effectively regulate the temperature field in the reaction zone in several variants: the radius of the apparatus and the speed of the flow of the reactants, the use of the band model of a rapid chemical process and the use of shell and tube apparatus with a bundle of small-radius intensification of convective heat transfer at profiling apparatus.

The process of neutralizing the oil in accordance with the proposed method is following (Fig. 13.3). The main flow of commercial oil from the pipeline (I) is mixed with the de-emulsifier (II), the pump (1) is directed to the heat exchanger (2), where it is heated to 110–120°C. The oil with de-emulsifier comes to the first stage of separation in electric dehydrators E1 (3). The oil from electric dehydrators (3) from top comes with the flow rate 680–750 m³/h at the inlet of the first section of a turbulent tubular reactor (4). Dispersion occurs in the five sections of the tubular turbulent reactor (Fig. 13.4), which is less than 4 meters with a pressure drop at the ends of the device to 0.52 atm. Aqueous alkaline solution (III) by pump (5) is sent to the coaxial connector of the first section of a turbulent tubular reactor (4) confusor-diffuser design with end nozzles (Fig. 13.5). Pipe is perforated by 20-one hole with a diameter $d_1 = 5$ mm, where 20 holes are in the walls for the radial outlet to the flow of oil supply bases, and closed front end of the pipe is perforated by hole coaxial with the direction of oil entering the solution of a neutralizing agent. Perforations are arranged symmetrically on the cross section (four holes on one section A-A). Partially dehydrated and desalted oil comes under pressure in the second stage of electric dehydrators E2 (6). Before this, electric dehydrators oil mixed in the diaphragm mixer (7) with preheated to 65–70°C pumped (8) fresh water (IV). Electric dehydrators E1 and E2 (3 and 6) by automatic reset valve saltwater (9) and (10) disperse the water to the sump (11). The extracted water streams containing oil are received for recycling by pumps (12) and (13). Desalted and dehydrated oil from the top V electric dehydrators E2 (6) play with the installation.

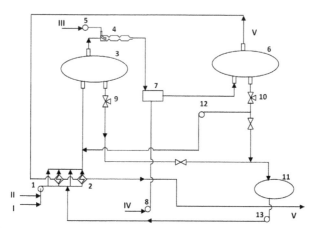

FIGURE 13.3   Scheme of the site electric desalting oil. 1, 5, 8, 12, 13 – pumps; 2 – heat exchanger; 3, 6 – electric dehydrators, 4 – turbulent tubular reactor, 7 – diaphragm mixer, 9, 10 – valve automatically reset the salt water, 11 – the sump.

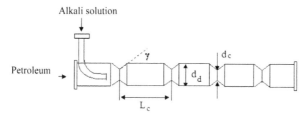

FIGURE 13.4   General view of the tubular turbulent apparatus for neutralization of the petroleum with alkali.

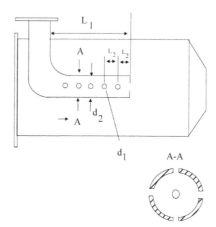

FIGURE 13.5   Scheme for input socket bases.

13.2   CONCLUSIONS

1. Tubular turbulent apparatus confusor-diffuser design allows for effective neutralization of aqueous alkali oil and organic amines in equimolar ratio.
2. The proposed low metal device confusor-diffuser design determines the differential pressure at the ends of the device of the five sections of no more than 0.52 atmospheres and is installed as part of the pipeline flow of oil on the node CDU.

This work was supported by the grant of the President of Russian Federation MD-3178.2011.8, RFBR (No. 11-03-97017).

KEYWORDS

- **Confusor-Diffuser Design**
- **Oil Neutralization**
- **Tubular Turbulent Apparatus**

REFERENCES

1. Sorochenko, V. F., Shutko, A. P., Pavlenko, N., & Bukolova, T. P. (1984). The effectiveness of corrosion inhibitors in water recycling systems, Chemistry and technology of fuels and oils, 7, 37.
2. Yushmanov, G. A., Starostin, N., & Dyakov, V.G. (1985).Current status of anti-corrosion protection techniques and material selection for equipment installation training of primary oil refining.CNIITpetrochemestry, Moscow,
3. Bergstein, N. V., Khutoryansky, F. M., & Levchenko, D.N. (1983).Improving the process for desalting EDU. Chemistry and Technology of fuels and oils 1, 8.
4. Khutoryansky, F. M., Zalischevsky, G. D., Voronin, N. A., & Urivskii, G. M. (2005). The use of a static mixer for enhanced mixing desalinated oil with an aqueous solution of alkali. Refining and neftehimiya, 1, 11.
5. Zalischevsky, G. D., Khutoryansky, F. M., Warsaw, O. M., Urivskii, G. M., & Voronin, N. A. (2000). Pilot plant evaluation of the static mixer SMV type of firm "Sulzer Hemiteh" with desalting by CDU. Refining and Petrochemicals, 5, 16.
6. Zakharov, V. P., & Shevlyakov, F. B. (2006).Longitudinal mixing in the flow of fast liquid-phase chemical reactions in the two-phase mixture. Journal of Applied Chemistry, 79(3), 410.

7. Zakharov, V. P., Mukhametzyanova, A. G., Takhavutdinov, R. G., Dyakonov, G. S., & Minsker, K. S. (2002). Creating homogeneous emulsions tubular turbulent apparatus diffuser confusor structure. Journal of Applied Chemistry, 75(9), 1462.
8. Minsker, K. S., Zakharov, V. P., Takhavutdinov, R. G., Dyakonov, G. S., & Berlin, A. A. (2001). Increase in the coefficient of turbulent diffusion in the reaction zone as a way to improve the technical and economic performance in the production of polymers. Journal of Applied Chemistry, 74(1), 87.

CHAPTER 14

PROCESSING OF BLACK SULFURIC ACID TO GIVE THE DESIRED PRODUCTS

T. V. SHARIPOV, A. G. MUSTAFIN, R. N. GIMAEV,
F. KH. KUDASHEVA, and A. D. BADIKOVA

CONTENTS

ABSTRACT

A complex method of alkylation waste processing, namely of a black sulfuric acid is presented. It includes its neutralization, organic and inorganic phase extraction, obtaining of ammonium sulfate from the inorganic part. A way of producing complex fertilizers on a black sulfuric acid and ammonium sulfate basis obtained from waste is presented. The organic waste bottom product is used for obtaining bituminic and fuel compositions as reagents for increasing oil products recovery and detergents. Utilization of organic residues is carried out by thermal splitting.

14.1   INTRODUCTION

Black sulfuric acid amounts to no less than 30% of the general number of sulfuric acid production waste obtained by sulfuric acid alkylation of isoparaffins by high octane gasoline alkenes [1] with the octane alkylate being 95. Up to 130 kg of black sulfuric acid is formed at obtaining a ton of alkylate, that is, the acid output makes 13% of high-octane gasoline produced [2]. Black sulfuric acid of alkylation process is a complex compound of black liquid consisting of no less than 85% of H_2SO_4, 10% max of organic substances (recalculated as carbon) and water. The organic black sulfuric acid component is presented by sulfonic acids, sulfonic ethers, naphthenic and aromatic hydrocarbons, resins and asphaltenes [3].

Nowadays the main method for black sulfuric acid regeneration is its high temperature splitting at the presence of reductant with sulfur dioxide used and sulfuric acid obtained. Gaseous fuel, coke, hydrogen sulfide, sulfur, high-boiling petroleum fractions were used as reductants. High temperature splitting process is carried out at 950–1200°C. As a rule maintaining of high temperature is achieved by fuel and sulfur containing feedstock combustion (hydrogen sulfide and sulfur). The given method is quite universal and provides for black sulfuric acid regeneration of petrochemicals with hydrocarbon residues. However, it is characterized by high corrosion activity and considerate operation and energy costs [4–7].

In literature extraction, hydrolysis, reconcentration and freezing out are described as black sulfuric acid regeneration methods [3, 8–10]. In the work of Ref. [11], black sulfuric acid regeneration is carried out by liquid sulfur dioxide with formation of an immiscible liquid-liquid system accompanied by passing of some organic substances to the sulfur dioxide phase. There exist some methods of black sulfuric acid regeneration [12, 13], which include its reaction with hydrogen peroxide, peracetic and per-propionic acids at high temperature (70–80°C) and air supply to the reaction zone. In this case organic substances are decomposing and the COD indicator decreases from 110,000 to 4200–4500. However, the methods stated above are ineffective. So we conducted the research work aimed at working out methods for processing black sulfuric acid with target products obtained.

14.2   EXPERIMENTAL PART

For laboratory investigations the following agents were used:

Black sulfuric acid of TC2121-02-33818158-99 as a sulfuric acid alkylation waste of the unit "UPR NOVOIL" JSC constituting 86.0% of H_2SO_4, 10% of organic part (recalculated as carbon), 0.15% of residue after calcination. Density is 1.67 g/cm^3.

Aqueous ammonia by GOST 3760-79

Technical ethyl, technical acetone, alcohol, propyl, butyl and isobutyl alcohol, tetrahydrofuran, 1-4 dioxane were used as extragents.

Laboratory experiments on neutralizing black sulfuric acid prediluted in ammonia water was carried out in a reactor at stirring until neutral reaction medium. Then the reaction mixture was extracted with separation of aqueous and organic phases. Organic and inorganic components of the initial black sulfuric acid were further processed with target products obtained.

The organic component of black sulfuric acid is complex including: sulfonic acids (37%), sulfonic ethers (8.4%), paraffin-naphthenic hydrocarbons (22.6%), mono- and polycyclic aromatic hydrocarbons (6.6% and 8.9%, respectively), resins (5.5%) and asphaltenes (11.0%). The organic component of black sulfuric acid is characterized by a wide boiling range and compounds with distillation temperature above250°C (Table 14.1).

TABLE 14.1   Fraction Structure of the Organic Part of Black Sulfuric Acid

Fraction structure	Mass fraction, %
Fraction 140–210°C	13–15
Fraction 210–250°C	14–17
Fraction 250–290°C	17–20
Fraction above 290°C	44–46
Losses	Up to 3

We suggest a complex method for black sulfuric acid processing which includes neutralization of the sulfuric acid, separation by extraction of organic and inorganic parts and their target application.

Ammonium sulfate [14] is offered to be obtained from black sulfuric acid by its neutralization in 20–25% ammonia water in the extragent presence and at 50–75°Ctemperature to pH achieving 6–8. Ethyl alcohol was used as an extragent at mass fraction of sulfuric acid (100% H_2SO_4,): extragent1:0.2–0.45. Reaction mass is spontaneously divided into organic (upper) and aqueous (lower) layers. The aqueous layer is a saturated ammonium sulfate solution, which is evaporated and dried to obtaining the target product. This method allows to get ammonium sulfate with few organic impurities (up to 0.1–0.5%) and use it as a nitrogen fertilizer.

To expand the range of raw components [15] we offered to use alcohol production waste as an extragent. It is a concentrate of ethanol head impurities of 70–96%vol. or an intermediate fraction of ethyl alcohol containing 1 sulfuric acid to 0.4–0.9 extragent mass.

Acetone; or ethyl alcohol and acetone mixture; or ethyl alcohol (acetone) with butyl, propyl, isobutyl or isoamyl alcohol; or ethyl alcohol (acetone) and tetrahydrofuran or 1,4-dioxane in the ratio: 1 sulfuric acid to 0.2–0.4 extragent mass are offered to increase the effectiveness of the organic part extracting process after black sulfuric acid neutralization. The ratio of the solvent mixture components for extracting organic impurities from black sulfuric acid is determined [16].

In the processes described above [14–16] a return to the solvent process obtained at distilling organic and aqueous phases is offered to carry out. The extragent recycling helps to reduce its consumption.

Complex fertilizers containing much sulfate sulfur possess increased agrochemical activity. Production of these fertilizers is based on neutralizing a mixture of phosphoric and sulfuric acids by ammonium and obtaining sulfoammophosphor pulp [17]. In this case phosphoric acid is preliminary mixed up with absorption liquid. Absorption of ammonium waste gases is carried out as a rule by aqueous solution of sulfuric acid. Absorption liquid is a saturated solution of ammonium sulfate. A technology for producing complex fertilizers is suggested [18], where ammonium sulfate obtained from black sulfuric acid is additionally mixed up with phosphoric acid. An ammonium sulfate solution is given in the ratio 0.06–0.18 of solution to 1of phosphoric acid. This method allows to decrease the consumption of technical sulfuric acid for 4–10% and in this case ammonium sulfate evaporation and drying stages are eliminated. The scheme of black sulfuric acid processing and nitrogen (ammonium sulfate) and complex fertilizers obtaining is presented in Fig. 14.1.

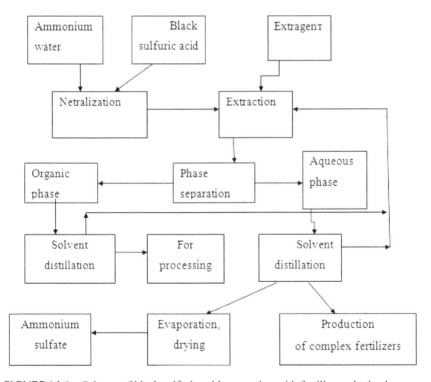

FIGURE 14.1   Scheme of black sulfuric acid processing with fertilizers obtained.

The technological scheme of processing black sulfuric acid with fertilizers obtained comprises the stages of acid neutralization, extraction, phase separation, distillation and solvent recycle and aqueous phase processing. Powdered ammonium sulfate obtained from black sulfuric acid may be used as a raw material for producing complex fertilizers. In this case ammonium sulfate is served on the granulation stage of a complex fertilizer. A method of direct using black sulfuric acid in producing fertilizers by obtaining diammonium phosphate [19] is worked out. It includes mixing of phosphoric acid and black sulfuric acid, neutralization of acid mixtures by ammonium, granulation, drying and further cleaning of waste gases in the antifoam (ПГ-3) presence. A 25–30 kg of black sulfuric acid is mixed with 1ton of sulfuric acid and then technical sulfuric acid is given up to 6.5–7.0% of SO_3 in the acid mixture. The given method allows to use black sulfuric acid and obtain dimed diammonium phosphate containing 18% of nitrogen and 46–47% of P_2O_5.

The isolated organic residue in obtaining ammonium sulfates from black sulfuric acid and after solvent distillation comprises sulfonate esters and acids, paraffin-naphthenic and aromatic hydrocarbons, resins and asphaltenes. It can be used as an adhesive additive for bitumen [20]. Bitumen is received by oxidation of heavy oil residue (tag) from the conventional viscosity to 200c at 80°Cwith the target product obtained and its further compounding with tar to receiving road asphalt. An organic residue amounting to 2–6 mas.% from the raw material is introduced into heavy oil residue before oxidation. To receive road asphalt compounding of the target product with tar is correlated in mas.%: BND brand 60/90: (80–75:20–25) and BND brand 90/130: (50–45:50–55). This method helps to improve plastic, low temperature and adhesive properties, bitumen heat-stability, shorten oxidation time and reduce the cost [21].

The presence of sulfuric acids with surface-active properties in the organic residue (up to 23% mass.) provides for its possible application in oil displacement, technical detergents and additives that increase the aggregate stability of fuel compositions.

A fuel composition [22] comprising 25–75% mass heavy pyrolysis resin, 22–74% mass furnace fuel oil, 0.5–3% organic residue of black sulfuric acid is characterized by high aggregate stability. Introduction of the organic residue is accompanied by the increase of colloidal stability neutralizing the fuel oil presence effect. The stability factor of the compositions ranges from 0.45 to 0.70. Fuel compositions with the organic component are recommended to be used as raw materials in technical carbon production.

Usage of the organic component of black sulfuric acid in compositions for oil displacement allows to diminish the surface tension on the "oil-composition" boundary from 28 mN/m to 4.50 mN/m and increase the oil recovery coefficient from 18% to 93%. The composition for increasing oil recovery [23] is a 2–10% aqueous solution of the organic residue.

The possibility of preparing detergent composition is experimentally tested using the organic residue of black sulfuric acid [24]. The pollution samples of technical carbon containing 13–22% of asphaltens, 15–61% of resins, 9–70% of paraffins and 1–23% of mechanical impurities are used as objects of the research. Technical detergents were received in the following way: the organic component was processed by 10% aqueous solution of caustic soda in mass fraction "organic component: NaOH" at 4:0,5concentration to 12–13pH at 60–70°C. Depending on the increase in paraffin content, mass loss of the pollutant decreases from 63% to 27% mass whereas dispersion grows from 28 to 61% mass at 10% mass of detergent concentration.

It should be noted that utilization of the organic residue of black sulfuric acid by its burning is possible in the existing units of thermal splitting of sulfuric acid. In this case corrosion of the whole process drops sharply providing for a long plant operation.

Thus, basing on the research made we showed the possibility of complex processing of black sulfuric acid by using its inorganic part for obtaining ammonium sulfate and fertilizers and by applying its organic residue for receiving a fuel composition, technical detergents as a special additive in bitumen, a reagent for increasing the oil recovery. Utilization of the organic part of the residue is carried out by its thermal splitting.

KEYWORDS

- **Ammonium Sulphate**
- **Bituminic Composition**
- **Black Sulfuric Acid**
- **Carbon**
- **Complex Fertilizer**
- **Detergent**
- **Extraction**
- **Fuel Composition**
- **Oil Product Recovery**

REFERENCES

1. Zh, Kirillova, V., Perfil'ev, V. M., & Sushev, V. S. (1984). Usage of Balck Sulfuric acids in USSR and abroad in Russian, M 30p.
2. Yu,V., Filatov, I. S., Lvova, G. A., & Eremina, T. M. (1972). Manelman Research in Regeneration and Usage of Black Sulfuric Acid in Russian, 22, M, 54.
3. Dorochinskiy, A. Z., Lyuter, A. V., & Volpova, E. G. (1970). Sulfuric acid Alkylation of Isoparaffins by Olefins in Russian M., "Khimiya" 216p.
4. Gimaev, R. N., Kondakov, D. I., & Syunyaev, Z. I. (1973). Modern methods of sulfur acid utilization of oil refinery and petro chemistry residues. In Russian Neftekhim, M., 97p.
5. Saenko, N. D., Agasieva, F. R., & Krasny, B. L. (2005). Pat.Rus. 2261219.
6. Lawrence, A., Smith, Abraham, & Gelbein, P. (2006). Pat. US 2006/0251570.
7. Lawrence, A., Smith, Abraham, & Gelbein, P. (2008). Pat. W. O 2008/030682.
8. Shenfeld, B. E., Vasilyev, B. T., Suschev, V. S., Zh, Kirillova, V., & Perfil'ev, V. M. (1986). Khim.Prom, 2, 97–99.
9. Dyumaev, K. M., Elbert, E. I., Sushev, V. S., & Perfil'ev, V.M. (1987). Regeneration of black sulfuric solutions, Khimiya, M., 112p.
10. Goncharenko, A. D., Perfil'ev, V. M., & Kostenko, A. S. (1982). Modern condition and perspectives of processing sulfuric acid waste, Nefterkhim, M., 52p.
11. William, M. (2010). Cross. Pat.US 2010/0152517.
12. Tse-Chaun Chou, & Yi-Lin, Chen. Pat. (1996). US 5547655.
13. Tse-Chaun Chou, Chao-Shan Chou, & Yi-Lin (1998).Chen. Pat. Canada 2194902.
14. Mustafin, A. G., Gimaev, R. N., Kh Kudasheva, F., Sharipov, T. V., & Badikova, A. D. (2008). Pat. Rus. 2325324.
15. Sharipov, T. V., Mustafin, A. G., & Kh Galiyanov, A. (2009). Pat. Rus. 2373149.
16. Sharipov, T. V., Mustafin, A. G., Kh Galiyanov, A., & Sh. Shayakhmetov, U. (2009). Pat. Rus. 2373150.
17. Sharipov, T. V., Mustafin, A. G., Volodin, P. N., & Usmanov, R. T. (2010). Pat. Rus. 2407727.
18. Mustafin, A. G., & Sharipov, T. V. (2011). Pat. Rus. 2411226.
19. Sharipov, T. V., & Mustafin, A. G. (2010). Pat. Rus. 2406713.
20. Sharipov, T. V., Mustafin, A. G., Kh. Galiyanov, A., & Evdokimova, N. G. (2010). Pat. Rus. 2405807.
21. Evdokimova, N. G., Mustafin, A. G., & Sharipov, T. V. (2012). Using black sulfuric acid regeneration of alkylation process as a modificatorin producing bitumen. Vestn. Bashk. Univers.1, ıovol.17, 42–47.
22. Kh Kudasheva, F., Sharipov, T. V., Gimaev, R. N., Badikova, A. D., Musina, A. M., & Mustafin, A. G. (2010). Pat. Rus. 2409613.
23. Gimaev, R. N., Kh Kudasheva, F., Badikova, A. D., Mustafin, A. G., Musina, A. M., & Sharipov, T. V. (2012). Pat. Rus. 2441049.
24. Kh, F., Kudasheva, A. D., Badikova, A. M., Musina, A., & Ya. Safina. (2010). Detergent composition from organic pollutants of chemical industries waste. Oil and gas industry http://www.ogbus.ru.

CHAPTER 15

STUDY OF CHROMIUM COMPLEXATION OF DIPHENYL CARBAZIDE ON ACTIVATED CARBON FIBER SURFACE

A. V. GRIGORYEVA, E. R. VALINUROVA, F. KH. KUDASHEVA, I. M. KAMALTDINOV, M. V. MAVLETOV, and I. SH. AHATOV

CONTENTS

ABSTRACT

The process of Cr (VI) complexation of modified carbon fibers was studied. Modified carbon fibers were obtained by the liquid-phase oxidation of the concentrated nitric acid and diphenyl carbazide solution impregnation. The morphology of starting and modified fibers surface was studied using atomic-force microscopy. Sizes of complexes formed in the process of interaction of chromium with the functional and analytical group of the diphenyl carbazide chelating reagent, was established.

15.1 INTRODUCTION

Carbon fiber materials are widely used in many industries because of their high sorption and kinetic features, electrical conductivity, thermal, chemical and radiation resistance. Using different methods of modification promotes obtaining new adsorbents with activity and selectivity for heavy metal ions [1, 2]. Liquid phase oxidation by strong acids leads to a considerable increase in acid sites of different strengths on the carbon fiber surface. Cation exchange properties of the oxidized fibers were mainly caused by carboxyl groups [3, 4]. The pH reducing of heavy metals salt solutions after the contact with oxidized fibers indicates the predominance of the ion adsorption mechanism [4]. However, adsorption of heavy metal ions is also accompanied by the surface complexes formation and changes in the transition metal oxidation states [5]. The modification of carbon materials by preadsorption of organic reagents with different functional and analytical groups on the adsorbent surface used to produce new adsorbents with specific properties [5]. A modifier on the support surface is fixed by the physical or chemical interactions. The latter are more preferable, as they allow obtaining significantly more stable sorbents.

Studies on the morphology of the surface of carbon fibers modified with organic chelating agents are currently insufficient, and so this topic is of great interest.

This chapter deals with the sorption of chromium with modified activated carbon fibers.

15.2   EXPERIMENTS

15.2.1   EXPERIMENTAL SET UP

The surface carbon fibers morphology was investigated by Agilent Technologies 5500 Scanning Probe Microscopy atomic force microscope. AFM survey was carried out with a semicontact method using a cantilever thickness of 4.0±1 mcm, length of 125±10 mcm, height of 30±7.5 mcm with the beam stiffness of 10–130 N/m at a resonance frequency of 204–497 kHz. The images are selected in the most informative presentations and processed using Pico Image Basic software. Before the study, samples of fibers were stuck to the double-sided tape and placed on a glass slide.

15.2.2   MATERIALS AND METHODS

Modified samples of carbon fibers were obtained with liquid phase oxidation by a concentrated nitric acid for 1 h and the treatment with 1% diphenyl carbazide solution. The initial carbon fiber used the activated carbon fiber brand УВИС-АК НПО "Uvikom."

The sorption activity of carbon fibers was studied under static conditions at room temperature and preselected pH range. The process solutions were prepared from a stock solution of potassium dichromate concentration of 1 g/L by their serial dilution by distilled water and acidified by the nitric acid. Adsorbent samples of 0.1 g were placed in a 50 mL flask in solutions with concentrations of 10 to 100 mg/L. Chromium content in the filtrates was determined by AAS on AA-6200 of Shimadzu PNDF 14.1:2.214-06 after the establishment of sorption equilibrium [6].

15.3   RESULTS AND DISCUSSION

The results of sorption of chromium from a potassium dichromate solution by starting and modified carbon fibers are shown in Table 15.1.

TABLE 15.1   Extraction Coefficients of Chromium Using Carbon Fibers

Adsorbent	R,%
ACF*	55
ACF$_{DPC}$	97
ACF+5 ml DPC 1%	100
OACF	75
OACF$_{DPC}$	95
OACF+5 ml DPC 1%	100

*ACF – starting carbon fiber, OACF – carbon fiber oxidized by nitric acid, ACF$_{DPC}$ and OACF$_{DPC}$ – starting and oxidized carbon fibers modified by diphenyl carbazide, ACF and OACF + 5 mL DPC – sorption of chromium using starting and oxidized carbon fibers in the presence of 1 mL of 5% diphenyl carbazide solution.

As follows from the data given, the oxidation of the carbon fiber has increased the level of chromium extraction for 20% whereas the preliminary applying to the diphenylcarbazide surface fibers raised it for 42%. But the maximum chromium adsorption takes place at the presence of the chelating agent. The effect observed shows that the chelating groups of the reagent-modificator more fully correspond with chromium in the solution. Then the complex formed is fixed on the carbon fiber surface with the help of aromatic diphenylcarbazide rings.

The surface state changes of carbon fibers after inoculation and adsorption of chromium was fixed using AFM.

The AFM image of the original activated carbon fiber at magnification of 5 × 5 μm² is shown in Fig. 15.1. From this image it is clear that nonmodified carbon fiber consists of a thin 7–8 micron diameter fibers composed of microfibrils 400–600 nm. The observed block structure fiber indicates the location of microfibrils along the fiber axis, in good agreement with literature data on the structure of carbon fibers [7].

Modification of carbon fiber with concentrated nitric acid helps to change the porosity of the fiber, as shown by the AFM image in Fig. 15.2. Pores with round and elongated shape are well seen in the picture. Liquid-phase oxidation of carbon fibers leads to significant increase in the meso- and macropores. Mesopore content of OACF increase to 60%, macropores – up to 22%.

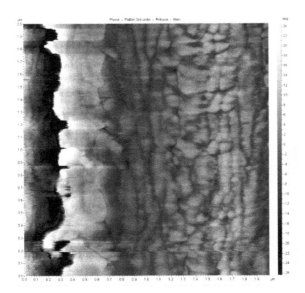

FIGURE 15.1   Phase (5 × 5 nm) AFM representation of the starting carbon fiber surface at different magnifications.

FIGURE 15.2   AFM image of the oxidized by nitric acid carbon fiber in the 2D format. Amplitude representation 2 × 2 μm.

Diphenyl carbazide coating to the surface of the oxidized carbon fiber is accompanied by the sorption latest in sorbent pores. Fiber surface is uniformly covered with molecules of chelating agent and gets rough form (Fig. 15.3).

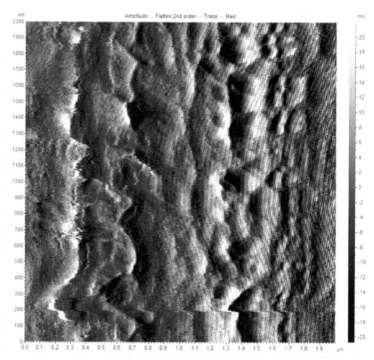

FIGURE 15.3 AFM image of 1% diphenyl carbazide solution carbon fiber modified with 1% diphenyl carbazide solution. Amplitude representation 2 × 2 μm.

The interaction of chromium with functional analytical group of diphenyl carbazide chelating reagent leads to the formation of the complex, its molecules are distributed evenly over the surface of the modified carbon fiber. Circular form complexes and their associates obtain the size of 50 to 100 nm (Fig. 15.4).

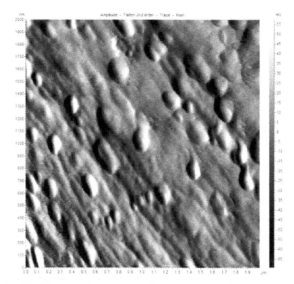

FIGURE 15.4 AFM image of the carbon fiber with complexes of chromium surface in the 2D form. Amplitude representation 5 × 5 μm.

15.4 CONCLUSIONS

Thus, it was established that modification of the carbon fibers by oxidation using nitric acid and diphenylcarbazide processing promotes efficient extraction of chromium from the water to 95–100%.

According to the AFM data oxidation of the carbon fibers improves the surface roughness and porosity.

Diphenylcarbazide premodification of carbon fibers leads to the formation of additional adsorption centers.

KEYWORDS

- **Adsorption**
- **Carbon Adsorbent**
- **Heavy Metal Ions**

REFERENCES

1. Tarkovskaya, I. A. (1981). The Oxidized Coal, Naukova Dumka, Kiev, 200p.
2. Zemskova, L. A., Avramenko, V. A., & Chernyh, V. V., etc. (2004). Journal of Applied Chemistry, 77, 1116–1119.
3. Gimaeva, A. R., Valinurova, E. R., Kudasheva, F. H., & Igdavletova, D. K. (2011). Sorption and Chromatographic Processes, 11, 350–356.
4. Varshavsky, V., Ya. (1994). Chemical Fibers Chemistry Moskow, 300.
5. Valinurova, E. R., Gimaeva, A. R., & Kudasheva, F. H. (2009). Vestnik Bashkirskogo Universiteta, 14, 385–388.
6. PND F 14.1:2.214–06 Quantitative Chemical Analysis of Water. Methods of Mass Concentration Measure of Ferrum Cadmium, Cobalt, Manganese, Nickel, Copper, Zinc, Chromium And Lead In Water and Sewage Probes by The Method of Plasma Atomic and Absorption Spectrophotometry.
7. Ermolenko, I. N., Lubliner, I. P., & Gulko, N. V. (1982). Carbon Fiber Materials Containing Elements. Science and Technology, Minsk, 27.

CHAPTER 16

BAND STRUCTURE OF SOLID SOLUTIONS OF COPPER AND SILVER CHALCOGENIDES

R. A. YAKSHIBAEV, A. D. DAVLETSHINA, N. N. BIKKULOVA, and G. R. AKMANOVA

CONTENTS

ABSTRACT

The band structure of solid solutions of copper and silver chalcogenides has been theoretically investigated by Quantum Espresso software package. The band structure is calculated for the high-temperature superionic cubic phase. The calculations showed that the transition from the low-symmetry phase into a high one is a change in the character of the chemical bond. Solid solutions AgCuX (X = S, Se, Te) in the high temperature phase are gapless semiconductors

16.1   INTRODUCTION

Semiconductor compounds $Me_{2-\delta}X$ (where Me = Cu, Ag and X = S, Se, Te) display a wide variety of physical and physical and chemical properties, that makes them promising for electronic equipment [1, 2]. A wide range of electrical and thermal properties, the ability to control these properties via controlled deviation from stoichiometry and high parameters of ion transport facilities make them convenient for the study of transport phenomena in mixed ionic and electronic conductors [3–5]. It should be noted that high cation conductivity and high value of self-diffusion coefficients are comparable with the conductivity and diffusion in liquid electrolytes and appear against the predominant electronic conductivity [6–8].

Phase transitions of binary chalcogenides depending on temperature and deviation from stoichiometry as well as their electrical properties are studied well enough [2].

The semiconductor dependence $\sigma_e(T)$ for stoichiometric compositions obtains the most common electrical property of binary chalcogenides. Deviation from stoichiometry influences the electrical conductivity value and even slight deviations from stoichiometry lead to degeneration of the electron gas. It also changes the semiconductor conductivity to the metal one. Conduction changes occur in some systems as a result of polymorphic transformations under temperature variations for fixed compositions [1].

Particular interest in the study of copper and silver chalcogenides is determined by the fact that in these systems cations exhibit abnormally high mobility. A number of studies attempt to explain the emergence of highly mobile cations by the structural features of the band, and the hybridization degree of the d-levels of the metal and p-levels of chalcogen in particular. This approach represents a higher level of the formation of the ion transport properties than a crystal chemical one and is considered quite promising. Therefore, along with the attempt to explain the peculiarities of the electron transport by the construction of the band structure, we are to pay attention to a possible impact of d-p hybridization on the cations delocalization and their high mobility.

The studies of the band structure of copper and silver chalcogenides have been conducted by various methods [9–12]. In work [9], the band structure of silver chalcogenides was calculated by the augmented plane wave method. Absolute magnitudes of the electron mass calculated agreed with the experimental data. Absolute values of the electron effective mass calculated for silver chalcogenides with perovskite structure, sodium chloride, and fluorite amount to nearly 0.1 m, irrespective of their structure, whereas experimentally determined effective electron mass ranges from 0.05 m to 0.24 m. Nonparabolicity of the conduction band near the bottom is proved by the s-s-hybridization effect. Calculations also show that the effective mass of the holes depends on the hybridization of 4d-states of Ag and p-states of the chalcogen, whereas the band gap width is dependent on ss- and pd-hybridization.

The band structure for Ag_2Te and Cu_2Te has been calculated by the linearized connected plane wave method [10, 11]. In work [10], the p-d-hybridization phenomenon in the electronic structure of silver telluride is examined, and the effective electron and hole masses are calculated, which amount to 0.039 m_0 and 1.3 m_0-2.1 m_0, respectively. It has been shown that a decrease in the degree of hybridization between the d bands of Ag and Te p-bands does not affect the band structure characteristics near the energy gap. Basing on the results obtained, the authors concluded that the p-d hybridization did not influence the dynamics of ions on α-Ag_2Te. The evaluation of Ag_2Te and Cu_2Te p-d hybridization is given in [11]. In Cu_2Te d-bands are more closely connected with the p-band as compared with Ag_2Te. Smaller degree of p-d hybridization results in a faster Ag diffusion of Ag_2Te than of Cu_2Te.

The study of the band structure for β-Ag_2S (monoclinic structure at room temperature) and α-Ag_2S (cubic structure above 453 K) is experimentally presented in [12] by using photoelectron spectroscopy and by using the FP-LMTO (full-potential linear muffin-tin orbital) calculation in theory. The band structure calculation was performed for a modified bcc structure, where silver atoms occupy two different positions: 1) four silver atoms occupy octahedral sites, 2) four silver atoms are distributed by octahedral and tetrahedral positions with equal probability. The distribution of silver atoms to octahedral and tetrahedral positions affects the band gap. If the atoms occupy only octahedral sites, the energy gap is missing. The band gap for low-temperature and high-temperature structures (atoms occupy tetrahedral and octahedral) amounts to 2 eV.

The band structure of the solid solutions has not been thoroughly investigated yet. In the compounds given the ionic conductivity is carried out by Ag and Cu cations [6, 7].

16.2   METHOD OF CALCULATION

The band structure calculation was carried out with the density functional theory using the pseudopotential plane wave basis of a Quantum Espresso software package [13]. The exchange-correlation effects were taken into account in the local density approximation (LDA). In this method calculations 3d-and 4 s-electrons are counted for Cu, 4d-and 5 s-electrons are accounted for Ag, and 3 s-, 3p-electrons for S, 4 s-, 4p-electrons for Se and 5 s-, 5p-electrons for Te, respectively. Ultra-pseudopotentials for Ag and Cu were used for calculations whereas pseudopotentials preserving the norm generated by the chalcogen program were applied. The kinetic energy of plane waves trimming amounted to 35 Ry (476 eV). The automatic selection of the reciprocal lattice points (k-points) along with the method of Monkhorst-Pack [14] on a grid of $4 \times 4 \times 4$ (with a shift of the origin) were used.

16.3   RESULTS

Solid solutions AgCuX (X = S, Se, Te) at above 473 K go to the superionic phase with a cubic structure [2, 15, 16]. This state is characterized by the cation disorder occupying tetrahedral positions in the fcc lattice of the chalcogen. The lattice parameters for the high-temperature phase of the compounds and atomic coordinates are listed in Table 16.1.

TABLE 16.1 Lattice Parameters of the Solid Solutions and Atomic Coordinates

Compound	Lattice parameter, Å	Atomic coordinates
AgCuS	5.72(9)	Ag (¼, ¼, ¼)
AgCuSe	6.08(5)	Cu (¾, ¾, ¾)
AgCuTe	6.37(5)	X (0, 0, 0)

LDA and PBE functional calculations provide for the similar results. The results of the band structure calculations are shown in Figs. 16.1– 16.3. In the right column the density of electron states is represented.

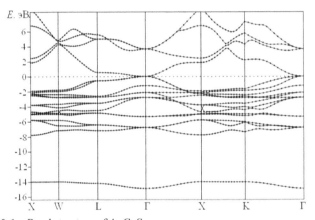

FIGURE 16.1 Band structure of AgCuS.

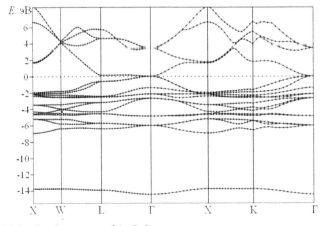

FIGURE 16.2 Band structure of AgCuSe.

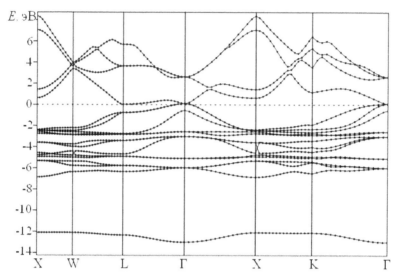

FIGURE 16.3   Band structure of AgCuTe.

Points of the Brillouin zone obtain the following coordinates in the units of the reciprocal lattice vectors: T (0, 0, 0), W (1/2, 0, 1), L (1/2, 1/2, 1/2), X (0, 0, 1), K (3/4, 0, 3/4). The last occupied state is taken for the zero energy. The energy level corresponding to −14 eV is formed by s-band with little contribution of chalcogen s-, d-levels of both types of metal and is not shown in Fig. 16.1. Within the range from −7.7 to the Fermi level there are p-states of sulfur and s-and d-states of silver and copper. In the interval from −7.7 eV to −5.7 eV the Ag d-states dominate whereas Cu d-states prevail in the range from −5 eV to the Fermi level.

Dependence of the energy from a wave vector in all cases is quite similar. In the band structure AgCuS (Fig. 16.1) there is an absence of the band gap characteristic of all the compounds studied.

In the cases of AgCuSe (Fig. 16.2) and AgCuTe (Fig. 16.3), a narrowing of the valence band compared to AgCuS is observed.

The bottom of the conduction band for all compounds is formed from the s-states of the cations and p-states of the anion.

An increase in the degree of metallicity of the chemical bond is indicated by the absence of the band gap in AgCuX systems for high-temperature phases. According to the electrophysical properties of these systems they are gapless semiconductors.

16.4 CONCLUSION

In this paper, the calculation of the band structure of the solid AgCuX solutions (X = S, Se, Te) was made. They exhibit high ionic conductivity in the high temperature cubic phase. The calculations showed that at high temperature the chemical bond of AgCuX (X = S, Se, Te) obtains a covalent-ionic character with a certain metallicity. The observed pd-hybridization results in the decrease of the effective cation radius, which causes their delocalization and mobility increase. Solid solutions AgCuX (X = S, Se, Te) in the high temperature phase are gapless semiconductors.

KEYWORDS

- **AgCuX systems**
- **Band structure**
- **Brillouin zone**
- **Copper and Silver Chalcogenides**

REFERENCES

1. Gorbachev, V. V. (1980). Semiconductor compounds $A_2^{I}B^{VI}$, Metallurgy, Moscow.
2. Semiconductor Chalcogenides, their alloys, (Ed). Abrikosova, N. H. (1975) Nauka, Moscow.
3. Kadrgulov, R. F., & Yakshibaev, R. A. (2001). Bulletin of the Bashkir University, 3, 13–14.
4. Kadrgulov, R. F., Yakshibaev, R. A., & Khasanov, M. A. (2001). Ionics 7(1–2), 156–160.
5. Yakshibaev, R. A., Konev, V. N., Mukhamadeeva, N. N., Balapanov, M., & Kh. (1988). Izv. USSR. Neorgan. Mater. T. 24(3) 501–503.
6. Miyatani, S. (1973). J. Phys. Soc. Jpn. 34 422.
7. Miyatani, S., Miura, J., & Ando, H. (1979). J. Phys. Soc. Jpn. 46 1825.
8. Yakshibaev, R. A., Kh. M., & Balapanov, V. N. (1987). Konev, Phys. 29, 937.
9. Hasegawa, A. (1985). Solid State Ionics, 15, 81.
10. Kikuchi, H., Iyetomi, H., & Hasegawa, A. (1997). J. Phys.Condens. Matter. 9, 6031–6048.
11. Kikuchi, H., Iyetomi, H., & Hasegawa, A. (1998). J. Phys.Condens. Matter. 10, 11439–11448.

12. Kashida, S., Watanabe, N., Hasegawa, T., Iida, H., Mori, M., & Savrasov, S. (2003). Solid State Ionics. 158, 167.

13. Quantum-ESPRESSO, http://www.quantum-espresso.org.

14. Monkhorst, H. J., & Pack, J. D. (1976). Phys. Rev. B. 13, 5188.

15. Yalverde, N. (1968). Z. Phys. Chea. к. F. 61, 92–107.

16. Nureyev, I. R., Salayev, E. Y., Nabiyev, R. N. (1983). Izv. USSR. Neorgan. Mater 19, 1074–1076.

CHAPTER 17

INTERACTION OF A PAIR OF IMPURITIES AND KINKS IN THE SINE-GORDON EQUATION

E. G. EKOMASOV and A. M. GUMEROV

CONTENTS

ABSTRACT

The kink dynamics of the sine-Gordon equation is studied in the model of the localized spatial modulation of the periodic potential. A case of two identical areas (or impurities) of the spatial modulation of the periodic potential is considered. It is shown that observing the collective effects of impurity influence is possible and depends on the distance between the impurities. A definite critical value of impurity distances causing two quite different ways of the dynamic kink behavior is demonstrated. The structure and properties of three-kink solutions of the sine-Gordon equation in the impurity area are studied.

17.1   INTRODUCTION

In recent years soliton dynamics attracts increasing researchers' attention [1]. It is explained by the fact that though solitons first appeared in studying integrable systems, they were applied in the nonintegrable systems soon describing many physical applications [2, 3]. For instance, solitons of the sine-Gordon equation in solid-state physics describe domain walls in magnets, dislocations in crystals, fluxions in Josephson junctions and contacts etc. In many cases the soliton behavior can be described by the point particle model and their time evolution may be complied with simple differential equations. However, perturbation influence account leads to considerable changes in the soliton structure that should be described as deformed particles [3]. At the same time inner degrees of soliton freedom may be excited and play a crucial role in physical processes. Inner modes include transmission and pulsation modes. The latter is connected with macrobiotic oscillations of the soliton width. The influence of various perturbation types on exciting the inner soliton modes in the sine-Gordon equation causes great interest. As the sine-Gordon equation describes many phenomena in different fields of physics it is quite clear that the solution of the equation in question turns to be quite urgent. For instance, many works are devoted to the study of time and coordinate influence on the external force described by delta-like, step, hyperbolic and harmonic functions [3–5]. If the research of small perturbation influence on the sine-Gordon equation solution can be carried out by the worked-out perturbation theory for solitons [2, 3, 6], the influence of big perturbations can be generally conducted by the help of numerical methods only [3, 7].

Spatial modulation of the periodic potential (or impurity) is of great interest as well [3, 6]. The problem of the sine-Gordon equation kinks and impurity interaction for the one-dimensional case has been long discussed in literature [6–9]. For example, the model of the classical particle for the kink and impurity interaction is applied in case the impurity is devoid of a mode as a localized vibrational state on the impurity [3]. Importance of impurity modes in kink and impurity interaction is demonstrated in the work of Ref. [9]. Much attention is drawn to multisoliton solutions of the sine-Gordon equation [10, 11].

Let us consider the modified sine-Gordon equation of the following type [6, 12]

$$(\partial^2\theta/\partial t^2)-(\partial^2\theta/\partial x^2)+(K(x)/2)\sin 2\theta=0, \tag{1}$$

where $K(x)$ is a function characterizing the spatial modulation of the periodic potential. In case $K(x)$ the Eq. (1) goes to the known sine-Gordon equation and is solved as a topological soliton or kink:

$$\theta(x,t) = 2\arctan(\exp[\Delta(\upsilon_0)(x-\upsilon_0 t)], \tag{2}$$

where $\Delta(\upsilon)=1/\sqrt{1-\upsilon^2}$, υ_0 is an uninterrupted parameter $(0<\upsilon_0<1)$ determining the kink velocity. There exists the spatial localized solution of the equation (1) as a stationary breather:

$$\theta_{breather}(x,t,\omega) = 2\ \arctan((\sqrt{1-\omega^2}\sin \omega t) / \omega \cosh (\sqrt{1-\omega^2}\ (x-x_0))) \tag{3}$$

where ω is the breather oscillation frequency and x_0 is its center coordinate.

There are multisoliton solutions of the sine-Gordon equation, for example, in Refs. [10, 11] an interesting three-kink solution of a wobble type is described:

$$f_{wobble}(x,t) = 4/\beta\ \arctan((\sqrt{1-\omega^2}/\omega)\sin (m\omega t)+1/2e^{\varepsilon m(x-x0)}(e^{-m}\sqrt{1-\omega^2}\ (x-x_0)+$$

$$+\rho^2 e^m(\sqrt{1-\omega^2}\ (x-x_0)))/(\cosh (m-\sqrt{1-\omega^2}\ (x-x_0))+ (\sqrt{1-\omega^2}/\omega)\rho e^{\varepsilon m}(x-x_0)\sin (m\omega t)) \tag{4}$$

where

$$\rho = (1-\varepsilon\sqrt{1-\omega^2})/(1+\varepsilon\sqrt{1-\omega^2}), \; -1<\omega<1, \; \varepsilon = +1 \qquad (5)$$

where x_0 is "the center" coordinate of the solution (however, as compared with the kink (2) this parameter does not coincide with the geometry kink center), ω – a wobble oscillation frequency. It is worth noting that β, m, ε parameters allow to change the general view of the solution.

The dynamics of the sine-Gordon equation solitons is thoroughly studied by analytical and numerical methods for the case of "point impurity," $K(x)=1-\varepsilon\delta(x)$, where $\delta(x)$ is a Dirac delta function, $0<\varepsilon<1$ [3]. As shown in case of approaching the "undeformed kink" the impurity acts as a potential. At the corresponding constant ε ($\varepsilon>0$) it acts as a magnetic potential so the sine-Gordon equation solitons may be localized and radiate. In case of approaching "the deformed kink" the deformation kink effects of a resonance character arise and are accompanied by the vibrational kink movement on the potential (e.g., strong changes in its form) [3]. The possibility of the impurity mode excitation at kink scattering influencing its dynamics was also taken into account. There was discovered kink reflection by the magnetic impurity, which was caused by resonance energy exchange between the transmission kink and impurity modes. A case of many point delta-like impurities interesting for some physical applications [13] and harmonic spatial modulation of the periodic potential [14] were also considered. The kink dynamics of the sine-Gordon equation $K(x)$ of a step type was studied by analytical and numerical methods.

For the case of spatially extended impurities

$$K(x) = \begin{cases} 1, \; x < x_1, x > x_1 + W \\ 1-\Delta K, \; x_1 < x < x_1 + W \end{cases} \qquad (6)$$

where W is the width of the spatial modulation of the periodic potential with its left boundary at x_1. The kink and impurity interaction for both undeformed and deformed kink models is researched [12, 15, 16]. The dependence of velocity and structure of the kink, soliton and breather from time is found and the minimum velocity necessary for the kink to overcome an impurity potential well is calculated. Influencing of the nonlinear wave of the single impurity on the kink dynamics is carried out numerically [15]. In Ref. [16] the possibility of resonance interaction of the kink and the excited impurity mode is revealed numerically and analytically and in this case the task is solved without a precise analysis of the kink structure at interacting with the impurity. The analysis of the structure and properties of the excited localized nonlinear waves on the impurity was carried out numerically in Ref. [12]. A case of the isotropic impurity [17] and nontrivial time metrics [18] was considered in the sine-Gordon model framework.

For the case of two identical impurities [19] the possibility of strong collective effects in the system was shown. In the given work possible dynamics of the sine-Gordon kinks in the model with two identical spatially extended impurities are studied taking into account possible excitement of localized high-amplitude nonlinear multisoliton waves.

17.2   INTERACTION OF THE KINK AND MAGNETIC IMPURITIES

Let us consider the case when the kink and impurity size are of the same order. The spatial modulation of the periodic potential is to be modeled as:

$$K(x)= 1, x<x_1, x_1+W<x<x_1+W+d, x>x_1+2W+d$$

$$1-\Delta K, x_1<x<x_1+W, x_1+W+d<x<x_1+2W+d \qquad (7)$$

That is, two identical impurities separated by a definite distance d. Obviously, at $\Delta K>0$, the impurity is a potential well for the moving kink whereas at $\Delta K<0$ it is a potential barrier.

The Eq. (1) was solved numerically using an explicit scheme. The equation discretization was carried out by the standard five-point scheme of the "cross" type with the stability condition $(\Delta t/\Delta h)^2<0.5$, where Δt is a time step and Δh is a coordinate step. Initially we have the kink (2) moving at a constant velocity υ_0, with boundary conditions of $\theta(-\infty,t)=0$, $\theta(+\infty,t)=\pi$, $\theta`(+\infty,t)=\pi$.

Common numerical realizations of the Eq. (1) used in Refs. [12, 15] enable us to calculate the kink structure and dynamics accurate enough to watch the kink pinning and passing through the impurities, structure and properties of the nonlinear waves excited. Yet much higher accuracy is required to study possible resonance effects. So $N_x=10^4$ points were used for approximating the function $\theta(x,t)$. Special control of the result errors is carried out.

In the course of the numerical experiment the kink crosses the impurities and the kink structure and its main dynamic characteristics at every moment of time are calculated. In Fig. 17.1, the found kink dynamics is demonstrated. Among them we obtain the following: a kink is taken in the first (Fig. 17.1, curve 2) or second impurity (Fig. 17.1, curve 3); the kink oscillates between them for some time (Fig. 17.1, curve 1), reflects in the reverse direction (Fig. 17.1, curve 5) or passes the impurity area (Fig. 17.1, curve 4). In the latter two cases oscillated localized high-amplitude nonlinear breather waves are excited on the impurities (Fig. 17.2), which considerably influence the kink dynamics. Firstly, a considerable part of the initial kink energy may be spent on their excitation. Secondly, further interaction with these localized waves forms the resonance effect mechanism (in case of single impurities such interactions lead to reflecting from the magnetic potential [3]). In the case when the impurities are situated near each other, the energy necessary for transition between them is little and the kink may oscillate between them for a long time.

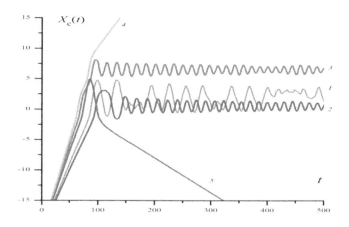

FIGURE 17.1 Dependence of the kink Xc center coordinate from time t for different modeling parameters at $\Delta K = 0.8$, $W = 1$: 1) the kink pinning on an impurity with a long jump [at Vo = 0.22, d = 1.5], 2) pinning on the first impurity [at Vo = 0.20, d = 5], 3) pinning on the second impurity [at Vo = 0.28, d = 5], 4) passing [at Vo = 0.30, d = 5], 5) reflection [at Vo = 0.26, d = 2].

In Fig. 17.3, a diagram of possible kink dynamics depending on the Vo and d parameters is presented and some critical distance between the impurities d_{crit} changing the system behavior is seen. In the given case $d_{crit} \approx 3.2$. At $d < d_{crit}$ two impurities may be treated as a single impurity. In Fig. 17.3 a horizontal line corresponds to the threshold velocity of a single defect v^{one}_{min} =0.245 and threshold velocities necessary for passing through the impurities either for single or double defect with the same parameters coincide at (d = 0.84).

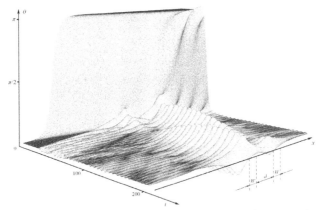

FIGURE 17.2 Dynamics of the kink passing through the impurity area and exciting high amplitude nonlinear localized waves of a breather type at $\Delta K = 1.2$, $W = 1$, $d = 5$, $Vo = 0.64$.

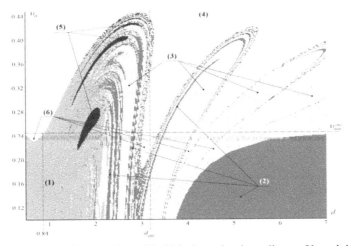

FIGURE 17.3 The diagram of possible kink dynamics depending on Vo and d at $W = 1$, $\Delta K = 0.8$. The diagram areas: (1) – kink pinning on an impurity with a long jump (2) – pinning on the first impurity; (3) – pinning on the second impurity; (4) – kink passing through both impurities area; (5) – full kink reflection; (6) – quasi tunneling area as a part of the area (4).

At $d>d_{crit}$ the diagram obtains "a petal" character: the pinning and passing areas start alternating. Such a behavior is caused by the fact that the kink energy losses spent on the breather exciting depend on its initial velocity. And when these losses exceed the definite value, the remaining kink energy is insufficient to break the magnetic potential of the double impurity and its pinning takes place. In Fig. 17.4, three vertical cuts of Fig. 17.3 at $d=5$ (Fig. 17.4a), $d=6$ (Fig. 17.4b), $d=7$ (Fig. 17.4c) are shown. It is found that in some particular cases the kink requires less kinetic energy for passing through two identical impurities of (7) type than for passing a single impurity of the same size. The area of initial kink velocities used to overcome the impurity regions (less than $v_0=v^{one}_{min} = 0.245$) may be called quasi tunneling (See Fig. 17.3, area 6). Note that earlier the same effect was obtained for the single impurity [18].

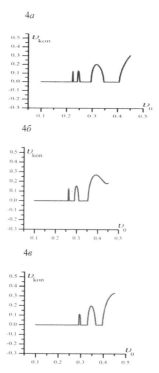

FIGURE 17.4   Dependence of the final kink velocity Vkon from the initial kink velocity Vo at $\Delta K = 0:8$, $W = 1$: a) $d = 5$, b) $d = 6$, c) $d = 7$. The cases Vkon $= 0$ correspond to the kink pinning on one of the impurities whereas Vkon > 0 comply with the kink passing through the double impurity area.

As follows from the results obtained, the critical distance d_{crit} divides the diagram into two parts characterized by strong and weak interaction of localized nonlinear waves. The properties of the excited breathers from the second diagram area (much differing from the typical kink dynamics at a single impurity) should be considered further.

17.3   EXCITING MULTISOLITONS ON IMPURITIES

We examine the case of the kink pinning on one of the impurities. Note that the presence of two impurity areas enables to find multisoliton solutions of the sine-Gordon equation. To be definite enough, we take $W = 1$, $\Delta K = 1.2$, $x_1 = -7$ and the distance d between the impurities may vary widely. In all cases hereinafter the initial kink velocity was selected so that the kink was pinned in the second impurity area. There are some difficulties in the numerical study of the task considered. As the interaction of the solitons excited may lead to oscillation mode appearance characterized by energy transfer from the kink to the breather and conversely (similar to the beat regime for harmonic oscillators), the oscillation frequency may change in the course of time. So further, we resort to isotropic oscillations and stationary frequencies that are set over time.

At the beginning we consider the case when the parameter d is high. As seen from Fig. 17.5a at $d = 10$ the kink is pinned in the second impurity area and transmission ω_{trans} and pulsation ω_{pulse} modes are excited whereas in the first impurity area the breather with $\omega_{breather}$ frequency is excited. The characteristic frequencies amount to $\omega_{trans} = 0.429$, $\omega_{pulse} = 0.875$, $\omega_{breather} = 0.92$ (Fig. 17.6). If we compare the values of the excited pulsation and transmission kink modes and oscillation breather frequencies for a single impurity [19], we see that they practically coincide. Thus, the soliton interaction in the impurity area is not observed at high d parameters. The obtained multisoliton consisting of the weakly interacting kink and breather is somehow connected with the well-known three-kink solution of nonlinear differential triton equations [1].

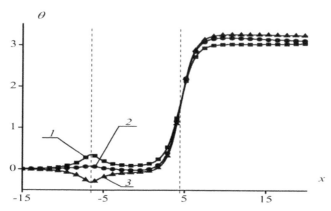

FIGURE 17.5 The multisoliton profile $\theta(x)$ in different moments of time: a) at $W = 1$, $\Delta K = 1$; 2, d = 10 in $t_1 = 1308.41$, $t_2 = 1309.91$, $t_3 = 1311.45$, б) at $W = 1$, $\Delta K = 1$; 2, d = 3 in $t_1 = 784.748$, $t_2 = 786.249$, $t_3 = 787.749$, в) at $W = 1$, $\Delta K = 1.2$, d = 1.6 in $t_1 = 1549.99$, $t_2 = 1551.49$, $t_3 = 1554.49$. Dashed lines denote center areas of the 1st and 2nd impurities.

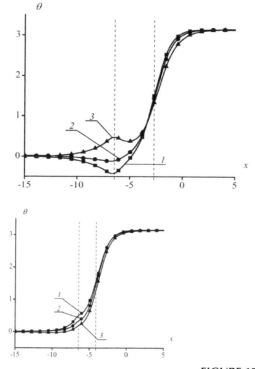

FIGURE 17.6 *(Continued)*

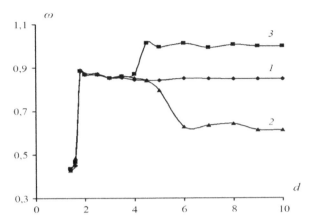

FIGURE 17.6   Dependence of the breather frequency $\omega_{breather}$ (curve 1), pulsation kink mode ω_{pulse} (curve 2) and transmission kink modes ω_{trans} (curve 3) from the parameter d at $W = 1$, $\Delta K = 1.2$.

Let us study the changes in the triton structure and properties at decreasing the d parameter. As seen from Fig. 17.5, we obtain a triton consisting of the noninteracting kink and breather at $d=5$ and above. Starting with d=4 (Fig. 17.5b), the triton structure and properties change greatly. The kink and breather are strongly bound and inner oscillation kink modes (pulsation and transmission) become equal to the breather oscillation frequency (Fig. 17.6). The multisoliton received can be considered as a certain triton solution of the sine-Gordon equation of the wobble type (4) at $\beta = 2$, $m = 1$, $\varepsilon = 1$.

In the third most narrow area the formation of a strongly bound kink and soliton condition is possible at $d \approx 1.4–1.6$ (Fig. 17.5c) that is a three-kink condition with much lower oscillation frequency $\omega_{pulse} \approx \omega_{trans} \approx \omega_{breather} \approx 0.46$ (Fig. 17.6). It corresponds to the results [19] for soliton oscillation frequency. For the triton case periodic energy transfer between the soliton and the kink is quite typical. The thing is that according to Ref. [12] the chosen parameters W and ΔK are to lead to the breather formation. Collective interaction effects of the impurities at a short distance result in the fact that a soliton is excited instead of the breather. Further decreasing the distance between the defects $d < 1.2$ makes it impossible to analyze the triton structure by the used numerical method.

17.4   CONCLUSION

The kink dynamics of the sine-Gordon equation was studied in the model with a localized spatial modulation of the periodic potential. As shown in the work, a collective effect of impurity influence strongly dependent of the distance is observed in the case of two identical areas of the spatial modulation of the periodic potential. The following ways of the kink dynamics were found: the kink is pinned in the impurity area and oscillates in between for a certain time; it is reflected in the reverse direction or passes the impurity area. In the latter two cases oscillated localized high amplitude nonlinear breather waves greatly influencing the kink energy are excited. Further interaction with the localized waves in question forms the basis for the resonance mechanism, namely "reflection from the magnetic impurity" and "quasi tunneling." There is found a definite critical distance value between the impurities that provides for two possible variants of dynamic behavior of the kink. In the first one the kink behavior is similar to the behavior in a single impurity. In the second case variants of the final kink behavior change depending on the initial kink velocity. It is explained by the breather oscillation phases excited in the second impurity area.

Pinning of the kink and exciting high amplitude localized nonlinear waves on the impurity may be used for multisoliton excitement in the sine-Gordon equation. A triton consisting of the weakly bound kink and breather is observed at long distances between the impurities. Starting with a certain critical distance pulsation and transmission mode frequencies are synchronized with the breather oscillation frequency and a triton solution of a wobble type is observed. At very short impurity distances excitement of the strongly bound kink and soliton is possible. The dependence of the structure and excited multisoliton frequencies from the impurity distances is determined.

KEYWORDS

- **Impurity**
- **Kink**
- **Multisoliton**
- **Sine-Gordon Equation**

REFERENCES

1. Encyclopedia of Nonlinear Science, Alwin Scott editor, Routledge New York and London, (2005), 1053p.
2. Shamsutdinov, M. A., Nazarov, V. N., Lomakina, I. U., & Kharisov, A. T. D. M.
3. Shamsutdinov Ferro & Antiferromagnitodinamika (2009). Nonlinear Oscillations, Waves and Solitons, Moscow, Nayka 456 p (in Russian).
4. Brown, O. M., & Kivshar, J. S. (2004). The Frenkel-Kontorova model, Concepts, Methods, and Applications Springer, 519 p.
5. Chacon, R., Bellorin, A., Gonzales, J. A. (2008). Phys. Rev. E. 77, 046212.
6. Yu, Kivshar, B., Malomed, F., Zhang, L., & Vazquez. (1991). Phys. Rev. B 43, 1098.
7. Fogel, M. B., Trullinger, S. E., Bishop, A. R., & Krumhandl, J. A. (1976). Phys. Rev. B., 15(3) 1578–1592.
8. Currie, J. P., Trullinger, S. E., Bishop, A. R., & Krumhandl, J. A. (1977). Phys. Rev. B., 15, 5567–5580.
9. Paul, D. I. (1979). J. Phys. C, Solid State Phys., 12(3), 585–593.
10. Kivshar, Yu, Zhang, F., & Vazquez, L. (1991). Phys. Rev. Lett. 67, 1177.
11. Kalberman, G. (2004). J. Phys. A., Math. Gen., 37, 11603–11612.
12. Ferreira, L. A., & Piette, B. (2008). Physical Review, E., 77, 036616.
13. Ekomasov, E. G., Azamatov, S. A., & Murtazin, R. R. (2008). The Physics of Metals and Metallography 105, 313–321.
14. Gulevich, D. R., & Kusmartsev, F. V. (2006). Phys. Rev. B., 74, 214303.
15. Gonz´alez, J. A., Cuenda, S. S. & Anchez, A. (2007). Phys. Rev. E., 75, 036611.
16. Ekomasov, E. G., & Shabalin, M. A. (2006). The Physics of Metals and Metallography, 101, suppl. 1, 48-S50.
17. Piette, B., & Zakrzewski, W. J. (2007). J. Phys. A Math. Theor. 40, 5995–2010.
18. Zhang, F., Kivshar, Yu, & Vazquez, L. (1992). Phys. Rev. A 45, 6019.
19. Yavidan, K. (2008). Phys. Rev. E 78, 046607.
20. Paul, D. I., (1978). Phys. Let. 64A(5), 485–488.

CHAPTER 18

VOLTAMMETRIC ELECTRONIC TONGUE APPLICATION FOR IDENTIFICATION OF MINERAL OILS

A. V. SIDELNIKOV, D. M. BIKMEEV, F. KH. KUDASHEVA, and
V. N. MAYSTRENKO

CONTENTS

ABSTRACT

For the first time is suggested that the analyte itself, that is, motor oil can be used as a binder into a carbon-paste electrodes. It is revealed that the electroactive markers: o-nitrophenol, o-nitrobenzoic acid, 2,5-dinitrophenol can specifically interact, accumulate and reduce on the CPE surface modified by different motor oils. Three-component mixtures of markers can be used for distinguishing motor oils with PCA applied. And their univocal identification is also possible by further using the SIMCA.

18.1   INTRODUCTION

Express methods of collecting information on the nature of motor and transmission oils and their characteristics are necessary for operative quality control of technique liquids in the process of exploiting. Another topical problem is monitoring of counterfeit products. Most of the present methods of motor oil quality evaluation require carrying out special analyzes such as sampling, sample preparation, determination and evaluation of major components and microimpurities of oils. Modern methods of analytical chemistry basing on chemometrics are able to decrease laboriousness and analysis time and extend a potential of analytical chemistry for identifying complex mixtures. The multisensor systems such as "an electronic tongue" being a combination of electrochemical sensors and chemometric data processing are of great use in monitoring alcohol products, juices, mineral drink waters, industrial liquids and etc. [1–6].

In our work an express voltammetric method of motor oils identification basing on carbon-paste electrodes (CPE) with motor oils as a binder is presented. It is revealed that the shape of voltammograms of electrochemically active markers reduction on CPE depends on the nature of motor oil. The registration of voltammograms of several markers for terms necessary for electronic tongues functioning [3]. Another ways of voltammograms registration, which can be used for motor, oils identification are a scan rate variation, changing the shape of polarizing voltage and etc. The responses can be formed both as an array of voltammograms of single marker reduction or as an array of voltammograms of marker mixture reduction.

In this chapter, the markers mixture which are able to reduce on CPE have been used for motor oils identification. A set of different motor oils of various natures (synthetic, semisynthetic and mineral) has been investigated.

18.2   EXPERIMENTS

18.2.1   EXPERIMENTAL SETUP

Carbon-paste electrodes were fabricated by mixing graphite powder (spectrally pure graphite) and motor oil in the ratio 6:1 (homogenization time 7–10 min). The fabricated paste was transferred into a glass tube with a diameter 2.0 mm. The surface of the electrode was renewed by removing 1–2 mm of paste after each voltammogram registration.

A standard three-electrode cell was used for registration of the analytical signal with the glassy carbon electrode as an auxiliary one, the Ag/AgCl electrode as a reference one, and carbon-paste electrodes on various oils basis as working ones.

Differential voltammograms of nitrocompound-markers (supporting electrolyte – 0.01 mol/L HCl) reduction were registered by using a voltammetric analyzer IVA-5 with the scan rate of 1 V/s, scan range from 0 to −1.5 V, after 15 s holding CPE in the standard solution of nitrocompounds with intensive mixing.

18.2.2   MATERIAL AND METHODS

$1 \cdot 10^{-4}$ mol/L solutions of *o*-nitrophenol (o-NP), *o*-nitrobenzoic acid (o-NB), 2,5-dinitrophenol (DNP) were used as standard solutions markers.

Mineral, semisynthetic and synthetic oils of different companies were investigated in Table 18.1 [7].

TABLE 18.1   Some Characteristics of Motor Oils

Symbol	Oil	SAE class	Density at 20°C, g/ml	Kinematic viscosity at 40 °C, cSt	Congelation point, °C	Flash-point, °C
Synthetic oils						
S1	Mobil	5W-40	0.850	91	−48	236
S2	Xado	5W-40	0.853	88	−42	225
S3	Shell	5W-40	0.853	72	−48	206

TABLE 18.1   *(Continued)*

Symbol	Oil	SAE class	Density at 20°C, g/ml	Kinematic viscosity at 40 °C, cSt	Congelation point, °C	Flash-point, °C
S4	Lukoil	5W-40	0.853	83	−41	230
S5	Ford	5W-30	0.852	85	−42	220
S6	Mobil	0W-40	0.850	75	−54	230
S7	Shell	0W-40	0.845	75	−45	222
S8	Castrol	10W-40	0.863	105	−33	200
Semisynthetic oils						
P9	Mobil	10W-40	0.870	98	−33	218
P10	Lukoil	10W-40	0.871	94	−34	219
P11	Castrol	10W-40	0.873	95	−36	189
P17	Shell	10W-40	0.880	92	−39	220
Mineral oils						
M12	Castrol	15W-40	0.883	106	−30	195
M13	Mobil	10W-40	0.875	90	−33	215
M14	Lukoil	10W-40	0.878	91	−33	217
M15	Shell	10W-40	0.874	95	−33	210

The experimental data array was formed by five parallel measurements, including 360 instantaneous current dI/dE values on differential voltammograms for each oil.

For data processing the principal component analysis was carried out [3, 4]. It allows converting voltammogram into a point at the principal components plane. The points of the samples with similar properties are clustered on this plane. The similarities and differences between the samples are observed by relative positions of the clusters.

For motor oils identification soft independent modeling of class analogy (SIMCA) was used [4]. The method is based on the assumption that all objects of the same class obtain similar properties and individual characteristics. When constructing the model only similarities must be considered, discarding such features as noise. To do this, each class of the calibration set is independently modeled by the principal component analysis. Belonging the sample of oil to the reference sample cluster was established by comparing the distance from the point of the tested voltammogram to the cluster with a standard deviation of points into the cluster and comparing the distance from the point of the tested voltammogram to the center of the reference sample cluster with a leverage [4]. The computer program The Unscrambler v8.0 was used for calculations.

18.3 RESULTS AND DISCUSSION

Comparing the voltammograms of markers mixture reduction (o-NP, o-NB, DNP) for different motor oils (Fig. 18.1) shows that the shape of voltammograms and the position of the peaks of markers reduction are depending on the nature of motor oil.

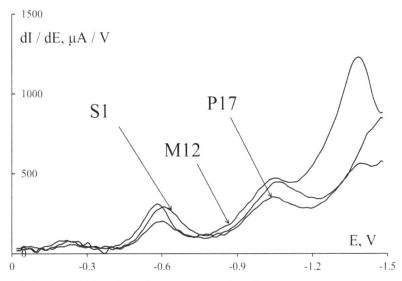

FIGURE 10.1 Differential voltammograms of markers mixture reduction on CPE modified by different motor oils (1×10^{-4} mol/L standard solution of a mixture of o-NP, o-NB, DNP, supporting electrolyte – 0.01 mol/L HCl solution, scan rate 1 V/s). S1 – synthetic motor oil Mobil 5W-40, P17 – semisynthetic motor oil Shell 10W-40, M12 – mineral motor oil Castrol 15W-40.

The scores plot of PCA-modeling of voltammograms of reduction of standard solution of a mixture of o-NP, o-NB, DNP on CPE modified by different motor oils is shown in Fig. 18.2.

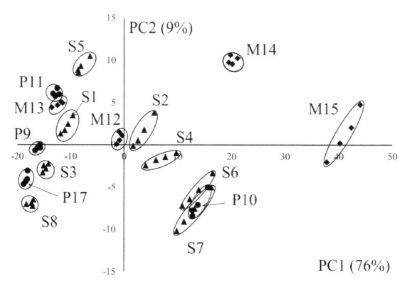

FIGURE 18.2   The scores plot of PCA-modeling of voltammograms of markers mixture reduction on CPE modified by different motor oils (1×10^{-4} mol/L standard solution of a mixture of o-NP, o-NB, DNP, supporting electrolyte – 0.01 mol/L HCl solution, scan rate 1 V/s).

PCA-processed voltammograms occupy separate clusters on the principal components plane depending on the nature of motor oil included into CPE as a binder. Herewith synthetic oils occupy mainly the second and third quarters of the principal components plane whereas the mineral ones mainly occupy the first and fourth.

Thereby the PCA-modeling scores plots can be used to reveal the information on the nature of motor oils and their subsequent identification.

The results of the identification of motor oils by use of standard 1×10^{-4} mol/L solution of a mixture of p-NA, o-NP and DNP with 0.01 mol/L HCl solution as supporting electrolyte are shown in Table 18.2. It is seen that using the standard solution of a mixture of three markers is useful for unambiguous identification of synthetic motor oils relative to mineral and semisynthetic ones.

TABLE 18.2 The Results of Motor Oils Identification ($1\ 10^{-4}$ mol/L standard solution of a mixture of o-NP, o-NB, DNP, supporting electrolyte – 0.01 mol/L HCl solution, scan rate 1 V/s).

CS*	Probability of identification, %							
TS**	P9	P10	P11	P17	M12	M13	M14	M15
S1	100	100	100	100	100	100	100	100
S2	100	100	100	100	100	100	100	100
S3	57	100	100	100	100	100	100	100
S5	100	100	100	100	100	100	100	100
S6	100	100	57	100	100	100	100	100
S7	100	50	100	100	100	100	100	100
S8	100	50	100	100	100	100	100	100

*CS: calibration sample, **TS: test sample.

18.4 CONCLUSIONS

1. For the first time is suggested that the analyte itself, that is, motor oil can be used as a binder into a carbon-paste electrodes.
2. It is revealed that the electroactive markers: o-nitrophenol, o-nitrobenzoic acid, 2,5-dinitrophenol can specifically interact, accumulate and reduce on the CPE surface modified by different motor oils.
3. Three-component mixtures of markers can be used for distinguishing motor oils with PCA applied. And their univocal identification is also possible by further using the SIMCA.

18.5 ACKNOWLEDGMENTS

The reported study was partially supported by RFBR, research project No. 11-03-00274-a.

KEYWORDS

- **Electronic Tongue**
- **Motor Oil**
- **Principal Component Analysis**

REFERENCES

1. Budnikov, G. K., Evtyugin, G. A., & Maystrenko, V. N. (2010). Modified Electrodes for Voltammetry in Chemistry, Biology and Medicine, BINOM. Knowledge laboratory, 416.
2. Maystrenko, V. N., Evtyugin, G. A., & Sidelnikov, A. V. (2011). Volta metric Electronic Tongue; Problems of Analytical Chemistry. Science, 14, 285–313.
3. Legin, A. V., Rudnitskaya, A. M., & Vlasov, Yu. G. (2011). Electronic tongue the Systems of Chemical Sensors for the Analysis of Aqueous Media. Problems of Analytical Chemistry. Science, 14, 79–126.
4. Rodionova, Y. O., & Pomerantsev, A. L. (2006). Russian Chemical Reviews, 75, 271–287.
5. Sidel'nikov, A. V., Zil'Berg, R. A., Kudasheva, Kh. F., Maistrenko, V. N., Yunusova, G. F., & Sapel'nikova, S. V. (2008). Journal of Analytical Chemistry, 63, 975–981.
6. Sidelnikov, A. V., Bikmeev, D. M., Kudasheva, F. Kh., & Maystrenko, V. N. (2013). Volta Metric Identification of Motor Oils Using an "Electronic Tongue." Journal of Analytical Chemistry, 68, 139–146.
7. URLs: http:www.mobil.com/. Accessed November 18, (2011), http:xado.ru/. Accessed November 18, (2011), http:www.shell.com.ru/. Accessed November 18, (2011); http:auto.potrebitel.ru/. Accessed November 18, (2011); http:www.castrol.com.ru/.

PART IV

BIOLOGICALLY ACTIVE SUBSTANCES SUCH AS
INHIBITORS OF THE CATALYTIC ACTIVITY OF
BIOPOLYMERS

CHAPTER 19

THE STUDY OF STRUCTURE–ANTIARRHYTHMIC ACTIVITY RELATIONSHIP OF BIOLOGICAL ACTIVITY COMPOUNDS USING SAR-METHODOLOGY

V. R. KHAIRULLINA, A. YA. GERCHIKOV, and F. S. ZARYDIY

CONTENTS

ABSTRACT

Using the pattern recognition theory methods relationships between the structure and antiarrhythmic activity of some nitrogen- and oxygen containing biologically active substances was been analyzed. The influence of the functional groups, as well as other structural descriptors on the antiarrhythmic properties has been evaluated. On base this descriptors set the model to predict antiarrhythmic activity of potential pharmaceutical preparations at the recognition level of 80% by means of two different approaches (geometric approach, the method of voting) has been generated

19.1 INTRODUCTION

Antiarrhythmic drugs are widely used in clinical practice. They differ by potency, mechanisms of antiarrhythmic action, and also by manifestations of side effects [1–3]. Many of currently used preparations are either insufficiently effective or exhibit the antiarrhythmic action during a long-term treatment [4, 5]. In some cases effective antiarrhythmic therapy (AAT) is even impossible for patients with ventricular arrhythmias due to fatal heart rhythm disturbances. Thus, a search for new biologically active compounds exhibiting high antiarrhythmic activity (AAA) still represents an important task.

The existence of relationship between structure of drug preparations and efficiency of their pharmacological action makes the program packages based of the "structure-property relationship" (SPR) applicable for searching new structures of potential drug candidates.

The aim of the present study was to investigate the "structure-AAA" relationship among N-phenylacetamide derivatives and aromatic carbonic acid amides and to apply the obtained data to the search of new potentially effective antiarrhythmic drugs.

19.2 MATERIALS AND METHODS

The "structure – AAA" study has been performed using the SARD-21 computer system (Structure Activity Relationship @ Design) [6]. This program determines structural descriptors responsible for manifestation of the activity of interest, quantitatively evaluates the degree of their influence, develops a mathematical model to predict this activity, and performs a complex of procedures on modification of known and generation of new structures to increase (decrease) the target property.

19.2.1 FORMATION OF A TRAINING SET

The training set was formed using literature data of structure and AAA activity of natural and synthetic derivatives of N-phenylacetamide and aromatic carbonic acid amides. It contained 41 structures of compounds exhibiting high antiarrhythmic activity (class A) and 63 structures of compounds, exhibiting rather low antiarrhythmic activity (class B). The ED_{50} values determined in the chloroform model on white albino mice were used as the classification criterion. The compounds with $ED_{50} \pounds 0.15$ mmol/kg were referred to the group of highly active compounds and those with $ED_{50} > 0.20$ mmol/kg were referred to the group of low active compounds [10–18]. These ED_{50} values have been chosen as limit criteria between classes of highly, moderately and low active compounds because quinidine included into the experimental set is generally considered as a moderately effective antiarrhythmic drug with the ED_{50} value of 0.22 ± 0.04 mmol/kg [19]. At the same time the drugs included into the experimental set, such as propafenone and disopyramide, exhibit a more pronounced antiarrhythmic effect and are characterized by the ED_{50} values of 0.07 ± 0.01 and 0.06 ± 0.01 mmol/kg, respectively [20, 21].

Table 19.1 shows typical structures of compounds included into the training set and corresponding ED_{50} values.

TABLE 19.1   Typical Structure of the Training Set

		ED_{50}, mmol/kg			ED_{50}, mmol/kg
Class of highly active compounds					
		0.14			0.08

TABLE 19.1 *(Continued)*

	ED_{50}, mmol/ kg		ED_{50}, mmol/ kg
	0.15		0.05
Class of low active compounds			
	0.23		0.70
	0.47		>0.50

19.2.2 PRESENTATION OF CHEMICAL STRUCTURE OF COMPOUNDS IN THE LEVEL OF FRAGMENTARY DESCRIPTORS IN TERMS OF FRAGMENTARY DESCRIPTORS (FD)

Three FD types were considered: 1) initial fragments including both elements of cyclic systems and cyclic systems themselves; 2) substructural descriptors of several chemically linked initial fragments; 3) logical combinations (conjunctions, disjunctions, strict disjunctions) generated on the basis of the first and second type descriptors.

19.2.3 EVALUATION OF INFORMATIVENESS OF ALL DESCRIPTORS

The mode of the FD effect on AAA was evaluated by means of the informativeness coefficient r (correlation of qualitative signs) ($-1 < r < 1$).

According to this coefficient the higher positive value of informativeness suggests higher probability of a given sign on manifestation of the analyzed property ((+) positive or (–) negative) [7].

19.2.4 GENERATION OF A MATHEMATICAL MODEL FOR RECOGNITION OF PREDICTION AND ITS TESTING BY MEANS OF COMPOUNDS OF KNOWN ANTI-INFLAMMATORY ACTION

Full descriptor characterization of compounds in two subsets containing 1700 descriptors is excessive. We have reduced the size of the descriptor characterization up to the optimal level and have determined the most significant factors: the decisive set of descriptors (DSD). The following criteria have been used for the sign inclusion into DSD: maximal informativeness, minimal interdependence, and optimal recognition of recognizable objects. Models for recognition and prognosis (prediction) were formed by combining DSD and classification rules as logical equations of the following type: C = F(S), where C is the feature (AAA), F is the recognition rule (an algorithm for image recognition, used to classify all the compounds under study; geometric method or the voting method), S is the set of recognizing decisive set of descriptors (DSD).

19.3 RESULTS AND DISCUSSION

We have generated DSD and the mathematical model for prognosis and recognition of AAA among N-phenylacetamide derivatives and aromatic carbonic acid amides (Table 19.2). An automatic selection within the algorithm has shown the fragmentary signs and their logic combinations potentially responsible for AAA have constituted DSD (Table 19.2).

TABLE 19.2 Decisive Set of Descriptors

No.	Property	Informativeness
1	{(-CH$_2$het-)-(CF$_3$)} # {(-NH-)-(1,2,4- trisubstituted benzene)} # {(CH$_2$het-(>N-)}	0.606
2	{(>C=O)-(1,2,4- trisubstituted benzene)} # {(-CHarom<)-(>N-)} #{(-CH$_2$)$_2$-)-(>N-)}	0.591

TABLE 19.2 *(Continued)*

No.	Property	Informativeness
3	{(-CHarom<)-(>C=O)} # {(-NH-)-(1,2,4- trisubstituted benzene)} # {(-CH$_2$het-)-(>N-)}	0.584
4	{(-CH$_2$het-)-(-O-)} # {(-NH-)-(1,2,4- trisubstituted benzene)} # {(-CH$_2$het-)-(>N-)}	0.549
5	{(-CHarom<)-(-O-)} # {(-O-)-(1,2,4,5- tetrasubstituted benzene)} #{(-CH$_2$het-)-(>N-)}	0.549
6	{(-O-)-(1,2,4- trisubstituted benzene)} # {(-CHarom<)-(>N-)} #{(-(CH$_2$)$_2$-)-(>N-)}	0.532
7	{(-NH-)-(1,2,4- trisubstituted benzene)} # {(-CHarom<)-(-O-)} #{(-CH$_2$het-)-(>N-)}	0.523
8	{(-CHarom<)-(-NH-)} ! {(-CH$_2$het-)-(>N-)} !{(-CHarom<)-(>N-)}	−0.575
9	(-NH$_2$) # (2,3- disubstituted naphthalene) # (7a-substituted hexahydro-1H-pyrrolo lysine)	−0.571
10	(1,2,4- trisubstituted benzene) # (>N-) # (-CHarom-)	−0.550
11	{(>CH-)-(>C=O)} # {(-CH$_2$het-)-(>C=O)} #{(-CH$_2$het-)-(>N-)}	−0.546
12	{(>CH-)-(-NH$_2$)} # {(-NH-)-(1,2,4- trisubstituted benzene)} #{(-CH$_2$het-)-(-NH$_2$)}	−0.530
13	{(-CH$_3$)-(-CHarom<)} # {(-NH-)-(1,2- disubstituted benzene)} #{(-CH$_2$het-)-(>N-)}	−0.523
14	(-CF$_3$) # (>N-) # (CHarom<)	−0.514

Note: **is** a conjunction sign (logical "and"), ! is a disjunction sign (logical "or and "), # is a sign of strict disjunction (logical "or no").

To validate the recognized dependencies the DSD testing was performed. We have used training and examination sets containing structures of known antiarrhythmic drugs. The maximal level of prediction of the target property was 85% for both training and examination sets evaluated using the voting method and the geometric approach, respectively (Table 19.3).

TABLE 19.3 Results of DSD Testing on Structure of the Examination Set

S. No.	Structural formula	Recognition by geometry	Recognition by voting	Literature data	
				ED_{50}, mmol/kg	Class of activity
1	 Quinidine	B	B	0.22	B
2		A	A	0.08	A
3	 Procainamide	B	B	1.30	B
4		B	B	0.58	B
5		B	B	0.35	B

6	Lidocaine	A	A	0.50	B
7	Disopyramide	A	A	0.06	A
8	Propafenone	A	A	0.07	A
9	Propranolol	A	A	0.10	A
10		B	B	0.10	A
11		B	B	0.54	B

While analyzing a sign space of the formed model the effect of individual groups and their various combinations has been evaluated. Trifluoromethyl group and tertiary nitrogen atom, and 1,2,4-trisubstituted benzene (among cyclic descriptors) are crucial for highly active antiarrhythmic drugs (Fig. 19.1, Table 19.4). Primary amino group and also mono-, 1,2-

and 1,2,3-trisubstituted benzene have been found mainly in low active compounds (Fig. 19.1, Table 19.4).

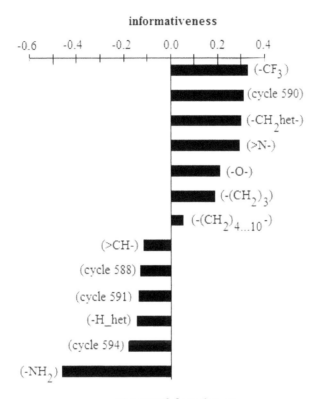

FIGURE 19.1   The effect of functional groups and cyclic fragments given in Table 19.4 on the efficiency of the antiarrhythmic effects of drugs.

TABLE 19.4   Cyclic Fragments and Their Corresponding Codes

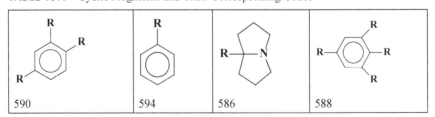

| 590 | 594 | 586 | 588 |

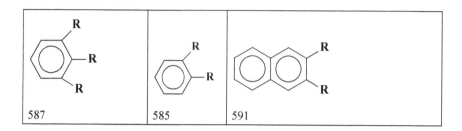

| 587 | 585 | 591 |

Analysis has shown that the degree and mode of the sign effect on manifestation of antiarrhythmic activity depends on nature and combination of adjacent descriptors. For example, sequential combination of secondary amino- and carbonyl group with 1,2,4-trisubstituted benzene fragment was typical for active compounds, whereas combination of the carbamide fragment with 1,2,4-trisubstituted benzene negatively influenced the target property (Fig. 19.2).

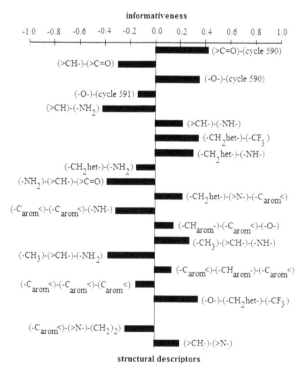

FIGURE 19.2 The effect of substructural descriptors on the efficiency of the antiarrhythmic effects of drugs.

19.4 CONCLUSIONS

1. Structure–AAA relationship of N-phenylacetamide derivatives and aromatic carbonic acid amides has been investigated, cyclic and acyclic signs typical for compounds with pronounced AAA have been recognized.
2. The decisive set of signs (DSD) that can predict existence of AAA in a wide spectrum of compounds and range them by the efficiency of the antiarrhythmic effect has been determined.
3. The mathematical model for prognosis of AAA with the recognition level of 82% by two approaches of the shape recognition theory (geometric approach, voting).

These data may be used for virtual screening of biologically active compounds to detect the presence of antiarrhythmic activity, AAA ranging, and also to determinate the targeted structural modification of known antiarrhythmic drugs to increase effectiveness of their action.

KEYWORDS

- **Antiarrhythmic Activity**
- **Shape Recognition Theory**
- **Structure-Activity**

REFERENCES

1. Sysoeva, N. A., (1993). Lechenie Narushenii Ritma Serdtsa (Treatment of Cardiac Arrhythmias), Moscow: Ovetlei.
2. Golitsyn, S. P. (1997). Klin Farmakol. Ter, 6(3), 14–16.
3. Rakhmanov, R. P. (2000). Osnovy antiaritmicheskoi terapii (Principles of Antiarrhythmic Therapy), Moscow, Anko.
4. Nedostup, A. V., & Blagova, O. V. (2003). Rus. Med. Zn, 11, 1168–1172.
5. Szabo, T., Geller, L., Mcrkcly, B., et al. (2000). Life Sciences, 66, 2527–2541.
6. Tyurina, L. A., Tyurina, O. V., & Kolbin, A. M. (2007). Metody i resultaty dizaina i prognoza biologicheski aktivnykh veschestv (Methods and Results of Design and Prediction of Biologicaly Active Substances), Ufa: Gilem.
7. Marot, C., Chavatte, P., & Lesieur, D. (2000). Quant. Struct. Act. Relat, 19, 127–134.
8. Kim, H. J., Chae, C. H., Yi, K. Y., et al. (2004). Bioorg Med Chem, 12, 1629–1641.

9. Poroikov, V. & Filimonov, D., (Eds) (2001). In Rational Approaches to Drug Design, Holtje, H. D., & Sippl, W., Barcelona, Porous Science, 403–407.

10. Sumoto, K., Satoh, F., & Shima, K. J. (1985). Med. Chem, 28, 714.

11. Banitt, E. H., Schmid, J. R., & Newmark, R. (1986). J. Med. Chem, 29, 299–302.

12. Banitt, E. H. & Coyne, W. E., (1980). J. Med. Chem., 23, 1130–1134.

13. Tenthorey, P. A., Adams, H. J., & Kronberg, G. (1981). J. Med. Chem., 24, 1059–1063.

14. Tenthorey, P., Block, A., & Ronfeld, R. (1981). J. Med. Chem., 24, 798–806.

15. Johnson, R., Baizman, R., & Becker, C. J. (1993). Med. Chem., 36, 3361–3370.

16. Campbell, K., Logan, R. T., & Marshall, R. (1986). J. Med. Chem., 29, 244–250.

17. Byrnes, E., Smith, E. R., & Mcmaster, P. (1979). J. Med. Chem., 22, 1171–1176.

18. Roufos, J., Hays, S. J., & Dooley, D. (1994). J. Med. Chem., 37, 268.

19. Jung, W., Andresen, D., & Block, M., et al. (2006). Clin. Res. Cardiol., 95, 696–708.

20. Reder-Hilz, B., Ullrich, M., & Ringel, M., et al. (2004). Naunyn-Schmiedeberg's Arch. Pharmacol., 369, 408–417.

21. Mergenthaler, J., Haverkamp, W., Hattenhofer, A., et al. (2001). Naunyn-Schmiedeberg's Arch. Pharmacol., 363, 472–480.

CHAPTER 20

RESEARCH OF MEDICINE STRUCTURAL DESCRIPTORS FOR TREATMENT OF MCF-7 DUCTAL INVACIVE BREAST CARCINOMA

V. R. KHAIRULLINA, A. YA. GERCHIKOV, M. N. VASILYEV, and F. S. ZARUDIY

CONTENTS

ABSTRACT

Using the SARD-21 system the structural descriptors characteristic for highly, medium and low effective antitumor compounds are revealed, the degree of their influence on malignant cells of invasive adenocarcinoma of ducts of mammary glands is estimated. Basing on the data obtained, the model of predicting interval levels of antitumor activity in sulfur-, nitrogen- and oxygen-containing heterocyclic compounds with the level of valid prediction of 80% by two methods of pattern recognition. The revealed regularities may be used for constructing new antitumor compounds

20.1 INTRODUCTION

Breast carcinoma (breast cancer) is much spread among women tumor diseases [1, 2]. Thus, the problem of searching medicines for treating it is quite topical. Modern researchers accumulated much information on effectiveness of growth inhibition of tumor cells in epithelial tissues by different classes of compounds, which in their turn are divided into alkylating agents, antimetabolites, antibiotics, and taxanes [3]. Much investigation in this field is determined by the fact that chemotherapy is proved to be effective for systemic treatment of patients with hormone-negative cancers whereas no considerable result is observed of patients with hormone-positive cancers [4, 5]. There are worked out new medical products acting as antioxidants and antimutagens able to cause apoptosis of tumor cells and restore the violated differentiation of epithelial and glandular tissue cells. They are aimed at restoring the structure and functions of healthy cells [6]. However, these data are quite uncoordinated whereas there is no information for studying the interrelation of the chemical structure and effectiveness of inhibiting the process of tumor development by different substances. Alongside with it there exists a real link between a chemical structure and pharmaceutical properties of many biologically active substances (BAS).

20.2 EXPERIMENTAL PART

The computer SARD-21 system (Structure Activity Relationship and Design) [7] was used for carrying out the investigations of the "structure-activity" relationship. Two models of predicting and defining interval levels of antitumor activity of natural and synthetic biologically active substances were constructed during the main procedures of the SARD-21 system.

The model M1 allows to reveal the structures of high, medium and low effective antitumor activity towards malignant cells of MCF-7 invasive carcinoma of ducts of mammary glands (adenocarcinoma). The typical compound structures of the learning set of the M1 model and the corresponding IC_{50} values are presented in Table 20.1.

TABLE 20.1   Typical Compound Structures of the M1 Learning Sets

The learning set of the M1 model is formed on the basis of 82 natural and synthetic heterocyclic nitrogen-, oxygen- and sulfur-containing biologically active compounds. They are divided into 2 groups with alternative properties: class A comprises 43 highly and medium effective antitumor compounds ($IC_{50} < 11$ µM for one of the abovementioned types of tumor cells) [1–6, 8–10]; class B includes 39 compounds with low antitumor activity ($IC_{50} > 16$ µM for one of the abovementioned types of tumor cells) [9–14].

The test set comprises 20 structures of highly, medium and low effective natural and synthetic biologically active substances characterized by high structural similarity of the training set structures of the M1 model (See Table 20.2).

TABLE 20.2 Results of Testing the Decisive Set of Signs of the M1 and M2 Models on the Structures of the Test Set

№	Structure	M1		Data in literary sources	
		Recognition by geom.	Recognition by vot.	In M1	IC_{50}, µM
1		A	A	A	0.11
2		B	A	A	0.004
3		B	A	A	4.1
4		A	A	A	8

5		A	A	A	0.023 ± 0.003
6	Epothilone A	B	A	B	5000
7		A	A	A	$0.19*10^{-9}$
8		A	A	A	12 ± 4
9		A	A	A	7
10		A	A	A	14
11		A	A	A	17 ± 4

12		A	A	A	13 ± 2
13		A	A	A	16 ± 3
14		A	A	A	14 ± 3
15		A	A	A	10 ± 3
16		A	A	A	16 ± 3
17		B	A	A	10.34
18		A	B	A	0.1

19		A	B	A	0.08
20	Tamoxifen	B	B	A	11.28

The structures of the researched chemical compounds were presented by fragmentary descriptors [7], namely: 1) initial fragments including cyclic systems and their elements; 2) substructural descriptors of some chemically related initial fragments; 3) logical combinations (conjunctions, disjunctions and strict disjunctions) generated on the basis of the first and second type of descriptors [12].

The influence of fragmentary descriptors on the antitumor activity was estimated with the help of the informativity coefficient r (correlation of qualitative features) ($-1 < r < 1$) in accordance with the higher positive value of informativity, the more the possibility of influencing the given sign on revealing the target sign (positive "+" and negative "–" ones) [7].

The model of recognition and prediction for the researched type of the activity was formed by the combination of classification rules and a number of structural parameters in the form of logical equations of C=F(S) type, where C is a sign (activity), F – rules of recognition (an algorithm of pattern recognition where the classification of the researched compounds is worked out, namely a geometric or "voting" approaches), S – a decisive set of descriptors. The effectiveness of the model of the researched activity type was determined by the results of the test set of the compounds and structures of the initial row using two methods of the recognition theory: a) geometric, and b) "voting" ones [7].

20.3   RESULTS AND DISCUSSION

Fragment descriptors and their logical combinations responsible for revealing the researched type of the activity were included in the decisive set of descriptors of the constructed model at the automatic selection of the algorithm under study (Table 20.3). In the M1 model the descriptors with a positive informativity coefficient are characteristic only for highly and medium effective compounds whereas a negative coefficient is typical for low tumor activity ones.

TABLE 20.3   Decisive Set of Signs for Constructing the M1 Model

Sign No.	Its content	r
1	(-NH-) ! 8,12-disubstituted-8H-benzo[b] pyrido [4,3,2-de]-1,7 phenanthroline -! 1,2,3,5- tetrasubstituted cyclohexane	0.836
2	{(-NH-)-(>C=C<)} ! {(>N-) - (>C=O)} ! {(>CH-)-1,2,3,5- tetrasubstituted cyclohexane}	0.712
3	{(-NH-)-(>C=O)} ! {(-NH-)-(>C=C<)} ! {(>CH-)-1,2,3,5- tetrasubstituted cyclohexane}	0.692
4	2,5- dihydro-2,3,4,5-tetrasubstituted furan # (>C=O) # (-(CH_2)_2-)	0.586
5	(>C=O) # (-(CH_2)_2-) # Cl	0.537
6	tetrahydro-2,3,4,5,6-pentasubstituted 2H-pyran # 6,9- disubstituted-9H-purine # (>C=O)	0.493
7	{(-CH_2het-)-(-O-)} # {(-CH_3)-(-O-)} # {(-CH_2het-)-(>C=O)	-0.685
8	(-O-) # 1,2,3,5- tetrasubstituted cyclohexane	-0.674
9	{(-CH_2het-)-{(>C=C<)} # {(>C<)-(-O-)} # {(-CH_2het-)-1,2,4-trisubstituted benzene}	-0.658
10	(>N-) # (F) # (>C<)	-0.634
11	1,2,4- trisubstituted benzene # monosubstituted benzene # (-OH)	-0.591

! – disjunction sign (logical "or")
– strict disjunction sign (logical "or not")

The results of recognizing the M1 learning and test sets using the decisive set of descriptors are given in Tables 20.2 and 20.4. These data testify that the constructed model obtains high recognition properties and can be used for predicting interval levels of antitumor activity of new compounds as its recognition level amounts to not less than 70%.

TABLE 20.4 Results of Recognizing the Learning and Test Sets Using the Decisive Set of Signs for the M1 and M2 Models

Recognition method	PH M1			
	Row A	Row B	Total	Test selection
Geom.	93.02	94.87	93.95	80.00
Vot.	74.36	84.62	79.49	80.00

Fragments characteristic for highly and medium effective antitumor compounds (the informativity coefficient $r \geq 0.1$) and descriptors of low effective antitumor compounds of the M1 model ($r \leq -0.1$) are revealed in the result of analyzing fragment descriptors of the constructed model. Taking into account their belonging to different functional groups the analysis of their influence is also fulfilled. For medium, low and highly effective antitumor compounds cyclic descriptors characteristic in Table 20.5 are presented.

TABLE 20.5 Cyclic Signs Characteristic for Highly, Medium and Low Effective Antitumor Compounds

536   543   544

548   549

*Fragment codes are marked by figures

*Fragment codes are marked by figures.

It is shown that the primary amino and nitrogroups (Fig. 20.1a) are both met among compounds of high, medium and low antitumor activity. Such descriptors as a carbonyl group, sulfur atom and secondary amino group are most frequent for highly and medium effective compounds (Fig. 20.1a). It is pointed out that halogens dubiously influence on the target type of the activity. In particular, a fluorine atom is mostly characteristic of compounds with high and medium antitumor activity whereas a chlorine atom is more frequently met among low effective compounds (Fig. 20.1a). Cyclic descriptors of compounds with a vivid activity type are characteristic of 1,2,3,5-tetra-substituted cyclohexane (cycle 105, Fig. 20.1a), 1,2,4-trisubstituted benzene (cycle 290, Fig. 20.1a), 1,2-disubstituted benzene (cycle 122, Fig. 20.1a), and 3,3,4,9-tetrasubstituted-2,3,4,9-tetragidronafto [2,3-b] thiophene (cycle 515, Fig. 20.1a). Compounds of low antitumor activity are characterized by such cyclic descriptors as 2,3,4,5,6-pentasubstituted tetrahydro-2H-pyran (cycle 544[1], Fig. 20.1a), 3,5,14,17-tetrasubstituted androstane (cycle 543, Fig. 20.1a), 1,4-disubstituted benzene (cycle 143, Fig. 20.1a) and monosubstituted benzene (cycle 102, Fig. 20.1a).

It is established that not only the nature of the fragments contained but also methods of their combination with the neighboring properties influence the effectiveness of the antitumor compound activity. The consecutive combination of the hydroxyl group with the nitro group and 1,2-disubstituted benzene (cycle 122, Fig. 20.1b) is characteristic for highly and medium effective compounds whereas the combination of the very functional groups with 1,4-disubstituted benzene (cycle 143, Fig. 20.1b) is mainly met in the class of compounds with low effective antitumor activity. Different combinations of the secondary amino group with a carbonyl fragment are characterized by different influence, namely, the combination of the secondary amino group with a carbonyl fragment is typical of highly and medium effective compounds. However, the combination of the carbamide fragment with 3,17-disubstituted androstane (cycle 548, Fig. 20.1b) is a distinguishing feature of low effective compounds. The influence of the tertiary amino group on the target activity may be commented in the same way.

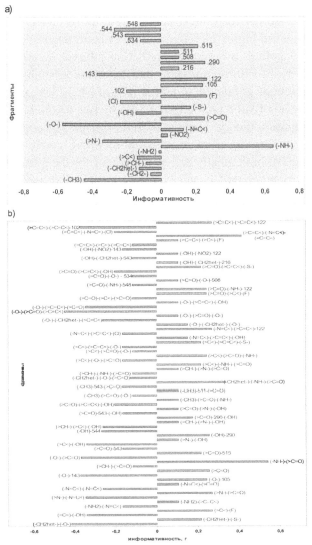

FIGURE 20.1 Acyclic fragments (Fig. 20.1a) and their combinations (Fig. 20.1b) characteristic for highly, medium and low effective antitumor compounds (structural formulas of the cyclic fragments are given in Table 20.5).

The results obtained may serve the basis for the directed synthesis of potential anticancer compounds.

20.4   CONCLUSION

1. There are revealed the structural descriptors characteristic for effective and highly effective antitumor compounds.
2. There is built a model of predicting interval levels of antitumor activity of sulfur-, nitrogen- and oxygen-containing heterocyclic compounds in relation to tumor cells of MCF-7 breast cancer with the level of valid prediction of 80% by two methods of the pattern recognition theory. The discovered principles of combining structural descriptors allow to carry out a virtual screening of different classes of biologically active substances for the presence of tumor activity in relation to cancer cells of invasive carcinoma of ducts of mammary glands (adenocarcinoma).

20.5   ACKNOWLEDGMENTS

The work has been carried out under the auspices of the Federal target program "Scientific and pedagogical staff of the Innovative Russia," State Contract No. 14.740.11.0367.

KEYWORDS

- Biologically Active Substances
- Breast cancer
- Built model
- Design and Synthesis
- Mechanisms of dietary
- System structural

REFERENCES

1. Chao, W. R., Yean, D., Amin, Kh., Green, C., & Jong, L. (2007). Computer-Aided Rational Drug Design, A Novel Agent (SR13668) designed to mimic the unique Anticancer Mechanisms of dictary indole-3-carbinol to Block Akt Signaling, J. Med. Chem, 50. 3412–3415.
2. Piacente, S., Masullo, M., De Ne`ve N., Dewelle, J., Hamed, A., Kiss, R., & Mijatovic, T. (2009). Cardenolides from Pergularia tomentosa display cytotoxic activity resulting from their potent inhibition of Na+/K+-ATPase. J.Nat.Prod, 72, 1087–1091.
3. Tardibono, L. P., Miller, Jr., & Marvin, J. (2009). Synthesis and Anticancer Activity of New Hydroxamic Acid Containing 1,4-benzodiazepines, Organic Letters, 11(7) 1575–1578.
4. Gomez-Monterrey, I., Santelli, G., Campiglia, P., Califano, D., Falasconi, F., Pisano, C., Vesci, L., Lama, T., Grieco, P., & Novellino, E. (2005). Synthesis and Cytotoxic Evaluation of Novel Spirohydanto in Derivatives of the dihydrothieno [2, 3-b]naphtho-4,9-dione system. J., Med. Chem., 48, 1152–1157.
5. Singh, P., Faridi, U., Srivastava, S., Kumar, J. K., Darokar, M. P., Luqman, S., Shanker, K., Chanotiya, Ch. S., Gupta, A., Gupta, M. M., & Negi, A. S. (2010). Design and Synthesis of C-ring Lactone and Lactam-based Podophyllotoxin Analogues as Anticancer Agents. Chem. Pharm. Bull, 58(2), 242–246.
6. Mu, F., Coffing, S. L., Riese, D. J., Geahlen, R. L., Verdier-Pinard, P., Hamel, E., Johnson, J., & Cushman, M. (2001). Design, Synthesis, and Biological Evaluation of a Series of Lavendustin a Analogues that Inhibit EGFR and Syk Tyrosine Kinases, as well as Tubulin Polymerization, J., Med. Chem., 44, 441–452.
7. Tyurina, L. A., Tyurina, O. V., & Kolbin, A. M. (2007). Methods and results of Projecting and Prognosticating Biologically Active Substances. Gilem, Ufa.
8. Delfourne, E., Darro, F., Bontemps-Subielos, N., Decaestecker, Ch., Bastide, J., Frydman, A., & Kiss, R. (2001). Synthesis and Characterization of the Antitumor Activities of Analogues of Meridine, a Marine Pyridoacridine Alkaloid, J. Med. Chem., 44. 3275–3282.
9. Meegan, M. J., Hughes, R. B., Lloyd, D. G., Williams, D. C., & Zisterer, D. M. (2001). Flexible estrogen receptor modulators, design, synthesis, and antagonistic effects in human MCF-/ breast cancer cells. J. Med. Chem., 44, 1072–1084.
10. Hakimelahi, Gh, H., Mei, N. W., Moosavi-Movahedi, A. A., Davari, H., Hakimelahi, Sh., King, K. Y., Hwu, J. R., & Wen, Y. Sh. (2001). Synthesis and Biological Evaluation of Purine-containing Butenolides, J. Med. Chem., 44, 1749–1757.
11. Son, J. K., Jung, S. J., Jung, J. H., Fang, Zh., Lee, Ch. S., Seo, Ch. S., Moon, D. Ch., Min, B. S., Kim, M. R., & Woo, M. H. (2008). Anticancer constituents from the roots of Rubia cordifolia L., Chem.Pharm.Bull, 56(2), 213–216.
12. Nicolaou, K. C., Pfefferkorn, J., Xu, J., Winssinger, N., Ohshima, T., Kim, S., Hosokawa, S., Vourloumis, D., Delft van, F., & Li, T. (1999). Total Synthesis and Chemical Biology of the Sarcodictyins. Chem. Pharm. Bull, 47(9), 1199–1213.
13. Jiang, M. M., Dai, Y., Gao, H., Zhang, X., Wang, G. H., He, J. Y., Hu, Q. Y., Zeng, J. Zh., Zhang, X. K. & Yao, X. Sh. (2008). Cardenolides from Antiaris toxicaria as potent selective Nur77 modulators. Chem.Pharm.Bull, 56(7), 1005–1008.

14. Reddy, M. V., Mallireddigari, M. R., Cosenza, S. C., Pallela, V. R., Iqbal, N. M., Robell, K. A., Kang, A. D., & Reddy, E. P. (2008). Design, Synthesis, and Biological Evaluation of (E)-Styrylbenzylsulfones as Novel Anticancer agents, J. Med. Chem., 51, 86–100.

CHAPTER 21

COMPUTER MODELING OF "STRUCTURE – ANTITUMOR ACTIVITY" RELATIONSHIP USING SAR-TECHNIQUES AND QSAR-METHODOLOGY

V. R. KHAYRULLINA, A. YA. GERCHIKOV, M. N. VASILIEV, I. A. TAIPOV, and F. S. ZARUDIY

CONTENTS

ABSTRACT

Some structural properties characteristics for highly, medium and low effective antitumor compounds are revealed using the computer system SARD-21 (Structure Activity Relationship and Design) and the degree of their influence on cancer cells of hepatoma, cervix and prostate is investigated. Basing on the data received two models of prognosticating interval levels of antitumor activity of sulfur, nitrogen, silicon and oxygen containing heterocyclic compounds with 80% reliability level are constructed by two recognition methods. The revealed structural regularities may be used for constructing highly active antitumor compounds. Additionally we have performed GUSAR (General Unrestricted Structure Activity Relationships) analysis of 30 quinazoline derivatives reported as thymidylate synthetase inhibitors. A robust, statistically sound and thoroughly validated consensus model is obtained. The six parametric model has the following statistical characteristics $N=30$, $R^2=0.926$, $F=34.093$, $SD=0.247$, $Q^2=0.884$, $V=6$. The GUSAR analysis provides the idea of regarding contribution of each atom in deciding the binding with thymidylate synthetase.

21.1 INTRODUCTION

Malignant tumors are one of the main reasons for high mortality rates [1]. In this regard the problem of searching medicines for treatment of these diseases is considered quite topical. It is known that the emergence and development of tumor cells in living organisms' tissues occurs due to uncontrolled processes of lipid peroxidation of their cell membranes [2]. And thus, a neoplasm which is non biocompatible with an organism's tissue emerges in the intercellular space.

Modern researchers accumulated quite a considerable information of effective inhibiting the growth of tumor cells of different etiology by various classes of compounds [3]. Many laboratories have at their disposal necessary medicines acting as antioxidants and antimutagens that are able to cause apoptosis of tumor cells and restore violated differentiation of liver, kidneys and brain cells. These drugs are aimed at preventing the tumor cells degeneration, at the same time restoring the functions of the healthy cells [4]. It is proved that application of such drugs does not cause any side effects in the body [5]. Many investigations are aimed at working out medicines of a new generation which obtain a more specific influence on tumor HepG2, HeLa, Hvr100-6, DU-145 cells (cytotoxicity) responsible for degenerating the liver, cervix and prostate correspondingly [6]. However, these data are quite uncoordinated whereas there is no information for studying the interrelation of the chemical structure and effectiveness of inhibiting the process of tumor development by different substances. Alongside with it there exists a real relationship between a chemical structure and pharmaceutical properties of many biologically active substances (BAS).

Thymidylate synthetase (EC 2.1.1.45) is the enzyme used to generate thymidine monophosphate (dTMP), which is subsequently phosphorylated to thymidine triphosphate for use in DNA synthesis and repair. Thymidylate synthase is induced by a transcription factor LSF/TFCP2 and LSF is an oncogene in hepatocellular carcinoma. LSF and thymidylate synthase plays significant role in liver cancer proliferation, progression and drug resistance. The enzyme is an important target for certain chemotherapeutic drugs [7].

This paper was focused on studying the "structure-antitumor activity" relationship of natural and synthetic biologically active substances to HepG2, HeLa, Hvr100-6 andDU-145 cells and thymidylate synthetase inhibitors using SAR-techniques and QSAR-methodology.

21.2   EXPERIMENTAL PART

The computer system SARD-21 (Structure Activity Relationship &Design) [8] was used for carrying out the investigations of the "structure-activity" link. Two models of prognosticating and defining interval levels of antitumor activity of natural and synthetic biologically active substances were constructed during the main procedures of the SARD-21 system.

The M1 model is aimed at identifying the structures with high antitumor activity towards Hep G2 hepatoma, HeLa and Hvr100-6 cervix and prostate DU-145 tumor cells. The M2 model allows to reveal medium and low efficient antitumor compounds. The training set of M1 model is formed on the basis of 108 natural and synthetic heterocyclic nitrogen-, silicium-, oxygen- and sulfur-containing biologically active compounds. They are divided into two groups with alternative properties: class A comprises 29 highly effective antitumor compounds (IC_{50}< 500 nM for one of the abovementioned types of tumor cells) [1–5]; class B includes 22 compounds with low antitumor activity (IC_{50} > 10^5 nM for one of the abovementioned types of tumor cells) and 57 compounds with medium antitumor activity (IC_{50} < 10^5 nM for one of the abovementioned types of tumor cells) [9–13]. The parameter IC_{50} as a boundary criterion for a high effective class of antitumor compounds from medium effective compounds separation was used. The training set includes 3',4'-didehydro-4'-deoxy-8'-norvinkaleykoblastin which is the acting compound of the cytostatic agent from the vinkaalkoloids group (IC_{50} = 376 nM for tumor cells of Hvr100-6 type, see Table 21.5), and compound (8S-cis)-8-acetyl-10-[(3-amino-2, 3,6-tridezoksi-alpha-L-liksogeksopiranozil)oxy]-7,8,9,10-tetrahydro-6,8,11-trihydroxy-1-methoxy-5,12-naftatsendion, that is the acting start of the cytostatic agent, daunorubicin (IC_{50} = 216 nM for tumor cells of Hvr100-6 type, see Table 21.5).

Model M2 comprises 57 medium-effective antitumor compounds (3300 <IC_{50}<10^5 nM, class A) and low-effective compounds with IC_{50}≥10^5 nM (class B) [14–15]. The value IC_{50} for a flavonoid baicalin, which has medium antitumor activity, served as a boundary criterion for the molecules of the M2 model training set between the classes of medium and low-effective compounds. The IC_{50} value of this substance amounts to 3330–16,300 nM for the above mentioned tumor cells.

The test selection comprises 21 structures of high-, medium- and low active natural and synthetic biologically active characterized by close structural similarity to structures of the training set of models M1 and M2 [3]. The typical structures in the training set of M1 and M2 models included compounds and their corresponding IC_{50} values are presented in Table 21.1.

The structures of the researched chemical compounds were presented by fragmentary descriptors [8], namely: i) initial fragments including both elements of cyclic systems and cyclic systems themselves; ii) substructures of several chemically linked initial fragments; iii) logical functions (conjunctions, disjunctions, strict disjunctions) based first and second type descriptors [13].

The influence of fragmentary descriptors on the antitumor activity was estimated with the help of the informativity coefficient r (correlation of qualitative features) (–1 <r< 1) in accordance with which the higher the positive value of informativity the more the possibility of influence of the given property on revealing the target property (positive "+" and negative "–" ones) is [8].

The models of recognition and prognosis for the researched type of the activity were formed by the combination of classification rules and a number of structural parameters in the form of logical equations of C=F(S) type, where C is a property (activity), F – rules of recognition (an algorithm of pattern recognition where the classification of the researched compounds is worked out, namely a geometric or "vote" method), S – a decisive number of properties. The effectiveness of the models of the researched activity types was determined by the results of the test set of the compounds and structures of the training set using two methods of the recognition theory of the samples: a) geometric, and b) "vote" ones [8].

The GUSAR program was used for quantitative research of relationships between the BAS structure and their thymidylate synthetase inhibition activity and QSAR models constrictions. It predicts the quantitative values of biological activity of chemical compounds on the basis of their "structural formulae" only and there is no need to have information about the 3D structure of ligands and target proteins. It has been verified that the GUSAR is a useful tool for QSAR modeling [16].

TABLE 21.1   Typical Structures of Compounds of the Training Arrays for Constructing M1 and M2 Models.

Prognosticatingandrecognizingmodelof highly effective antitumor compounds (M1)

Class of active compounds
(highly effective antitumor compounds)

IC_{50}= 2.94 ± 0.535 nM IC_{50}= 14.9 ± 4.28 nM IC_{50}= 198 ± 26.6 nM IC_{50}= 0.108 ± 0.0911 nM

IC_{50}= 15.0 nM IC_{50}= 36.6 ± 5.8 nM IC_{50}= 53.3 ± 11.5 nM IC_{50}= 20.9 ± 5.76 nM

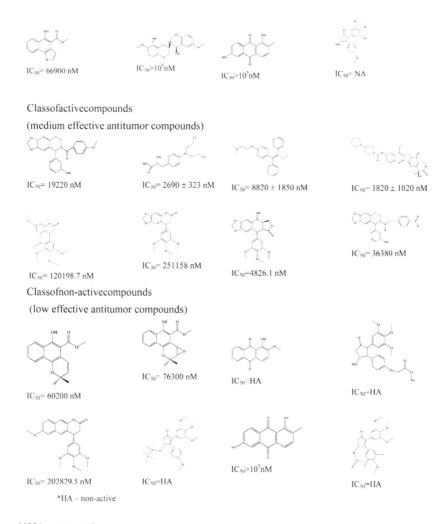

IC$_{50}$= 66900 nM

IC$_{50}$>10^{5}nM

IC$_{50}$>10^{5}nM

IC$_{50}$= NA

Classofactivecompounds

(medium effective antitumor compounds)

IC$_{50}$= 19220 nM

IC$_{50}$= 2690 ± 323 nM

IC$_{50}$= 8820 ± 1850 nM

IC$_{50}$= 1820 ± 1020 nM

IC$_{50}$= 120198.7 nM

IC$_{50}$= 251158 nM

IC$_{50}$=4826.1 nM

IC$_{50}$= 36380 nM

Classofnon-activecompounds

(low effective antitumor compounds)

IC$_{50}$= 60200 nM

IC$_{50}$= 76300 nM

IC$_{50}$=HA

IC$_{50}$=HA

IC$_{50}$= 202829.5 nM

IC$_{50}$=HA

IC$_{50}$>10^{5}nM

IC$_{50}$=HA

*HA – non-active

1*HA – non-active

A set of 30 quinazoline derivatives reported as thymidylate synthetase inhibitors [17] was used to test the performance of the GUSAR in QSAR. The set consists of diverse substituents from electron donating to electron withdrawing groups located at several positions in the bicyclic and acyclic core as shown in Table 21.6.

The 30 molecules were drawn in Marvin Scatch 5.9.1 freeware and biologic data addition before further analysis in GUSAR. For better analysis, following setting were used: Leave Many Out (LMO) = 20 iterations, Number of leave out = 20%, leverages = 0.99, Similarity=0.70, kNN RMSE/Average RMSE = 1, Number of Models = 20.

21.3   RESULTS AND DISCUSSION

Fragment properties and their logical combinations responsible for revealing the researched type of the activity were included in the decisive number of properties of each constructed model at the automatic selection of the algorithm under study (Tables 21.2–21.3). In the model M1 the properties with a positive informativity coefficient are characteristic only for highly effective compounds whereas a negative coefficient is typical for medium and low tumor activity ones. According to the decisive number of properties a positive value of the informativity coefficient for the model M2 corresponds to the properties of compounds with medium cytostatic activity. The properties with a negative coefficient value are characteristic for low effective tumor compounds.

TABLE 21.2   Number of Properties for Constructing M1 Model

Property Number	Its Content	r
1	{(>CH-)-(>C<)} ! {(-NH)-(>C=C<)} !{(-CH$_3$)-(>CH-)}	0.754
2	{(-CH$_2$het-) - (OH)} !{(-NH)-(>C=C<)} !{(-CH$_2$)-(>CH-)}	0.681
3	(-NH) # 2,3,4,6-tetrasubstituted tetrahydro-2H-pyran # 2,3,4,5,6-pentasubstituted tetrahydro-2H-pyran	0.660
4	{(-NH)-(>C=C<)} ! {(-CH$_2$-))-! {(>N-) - (>C=C<)}	0.631
5	{(-CH$_2$het-)-(>C<)} ! {(>C<)-(-O-) !{(-CH$_2$het-)-(-NH-)}	0.571
6	(-CH$_2$-) # 2,3,4,5,5-pentasubstituted −1,3-oxazolidine # (-S-)	0.553
7	1,2,3,5-tetrasubstituted benzene ! m-xylene! toluene	−0.548
8	m-xylene # 3,4,6-trisubstituted-1,2,4-triazinan # (-(CH$_2$)$_4$···$_{10}$-)	−0.482
9	5,6-disubstituted-5,6,7,8-tetrahydro [1,3] dioxol [4,5,9] isoquinoline# 3,4,6-trisubstituted-1,2,4-triazinan # (-(CH$_2$)$_4$···$_{10}$-)	−0.472

10	2,3,4,5,6-pentasusbtituted tetrahydro-2H-pyran # (-S-) # (>C=C<)	−0.432
11	3,4,6-trisubstituted-1 ,2,4-triazinan! toluene !(-(CH$_2$)$_4$...$_{10}$-)	−0.405
12	{(-O-)- (>C=C<)-(-O-)}	−0.330
13	(>N-) # (-OH) # (>CH-)	−0.326
14	p-xylene! (>C=O) ! (-OH)	−0.306

TABLE 21.3   Numbers of Properties for Constructing M2 Model

Property Number	Its content	r
1	1,2,3,5-tetrasubstituted benzene# (-NH-) # (-CH$_2$het-)	0.546
2	Si # -O- # -OH	0.492
3	{(-CH$_2$het-)-(>C=C<)}	0.400
4	{(-CH$_2$het-)-(-O-)}	0.340
5	2,4,5,6,7-pentasubstituted-4H-chromene # (-(CH$_2$)$_4$...$_{10}$-) # (-CH$_2$-)	0.301
6	{(>C=C<)-3,4,6-trisubstituted-1 ,2,4-triazinan } # {(CH$_3$)-(>CH-)} # {(>CH-)-(>C<)}	−0.793
7	{(1,2,3,5-tetrasubstituted benzene) - (3,4,6-trisubstituted-1 ,2,4-triazinan)} # {(-CH$_3$))-(>CH-)} # {(>CH-)-(>C<)}	−0.763
8	{(>C=C<)-(1,2,3,5-tetrasubstituted benzene)} # {(-CH$_3$)-(>CH-)} # {(>CH-)-(>C<)}	−0.763
9	{(-O-)-3,4,6-trisubstituted-1,2,4-triazinan)} # {(-OH)-3,4,6-trisubstituted-1 ,2,4-triazinan} # {(-NH)- (>C=O)}	−0.700

& – conjunction sign (logical "and")
! – disjunction sign (logical "or")
– strict disjunction sign (logical "or not")

TABLE 21.4   Results of Recognizing Training and Test Arrays Using the Decisive Number of Properties (DNP) Formed for M1 and M2 Models

Methods of recognition	DNP for M1 model				DNP for M2 model			
	Row A	Row B	Weight	Test selection	Row A	Row B	All array	Test selection
Geom.	82.76	93.33	85.05	70.00	98.15	76.00	87.07	88.00
Vote.	79.31	93.33	86.32	70.00	96.30	72.00	84.15	81.00

The results of recognizing the training and test sets using the decisive number of properties for M1 and M2 models are given in Table 21.4. These data testify both the fact of high recognition abilities of the constructed M1 and M2 models and the possibility of using them for prognosticating the interval levels of antitumor activity for new compounds as their recognition level is no less than 70%. A fragment of the training set and results of recognizing structures are given as an example in Table 21.5.

TABLE 21.5 Results of Testing the Decisive Number of Properties for M1 and M2 Models on the Structures of the Test Selection

| № | Structural formulum | M1 | | M2 | | Data in books | | |
		Recogniti on by geom.	Recogn ition by vot.	Recogniti on by geom.	Recogniti on by vot.	По M1	По M2	IC$_{50}$, nM
1	Daunorubicin	A	A	A	A	A	A	1.09 ± 0.31 (cells of HeLa line) 220 ± 90 (cells of Hvr100-6line)
2	Paclitaxel (Taksol)	A	A	A	A	A	A	0.136 ± 0.038 (HeLa cells) 562 ± 255 (Hvr100-6 cells)
3	Aklarubitsin	A	A	A	A	A	A	0.02 ± 0.01 (HeLa cells) 1.64 ± 0.588 (cells of Hvr100-6 line)
4	Vindesine	A	A	A	A	A	A	0.20 ± 0.1 (HeLa cells) 230 ± 50 (Hvr100-6 cells)

5	Actinomycin D	A	A	B	B	A	A	$(3 \pm 1) \cdot 10^{-4}$ (HeLa cells) 12.0 $\pm$ 1.4 (Hvr100-6cells)
6	SN-38	A	A	A	A	A	A	30.0 $\pm$ 5.0 (cells ofHeLa line) 28.3 $\pm$ 15.7 (Hvr100-6 cells)
7	Vinorelbine	A	A	A	B	A	A	$(6 \pm 3) \cdot 10^{-4}$ (HeLa cells) 380 $\pm$ 130 (Hvr100-6 cells)
8	Colchicine	A	A	A	A	A	A	-
9	baykalein	B	B	A	A	B	A	3300 - 16300 (DU-14 cells5)
10		B	B	A	A	B	A	$1.16 \cdot 10^{4}$ (HepG2 cells)
11	(-)-Talhonan-tluslakton	A	A	A	A	B	A	-

12	R(-)-Massoylakton	B	B	A	A	B	A	-
13	Monastrol	A	A	A	A	B	A	2.0•10⁴ (HepG2 cells)
14	Fluorouracil	A	A	A	A	B	A	(2.0 ± 0.5)•10³ (HeLa cells) 11500 ± 23300 (Hvr100-6 cells)
15	Nimustine	A	A	A	A	B	A	6.1•10⁷ (HeLa cells) 11300 ± 10300 (Hvr100-6 cells)
16	Cyclophosphamide	A	A	A	A	B	B	(1.7 ± 0.2)•10⁶ (HeLa cells) (1.4 ± 0.4)•10⁶ (Hvr100-6cells)

In the result of the joint analysis of fragmentary descriptors of both models there are revealed some fragments of the M1 model characteristic for highly effective antitumor compounds with the informativity coefficient $r \geq 0.1$ and properties of medium and low effective antitumor compounds of the model M2 with the informativity coefficient $r \leq -0.1$. The analysis of their influence is fulfilled taking into account their belonging to different functional groups. The cyclic properties characteristic for medium, low and highly effective antitumor compounds are presented in Table 21.7.

TABLE 21.6   Structure of High and Low Effective Thymidylate Synthetase Inhibitors.

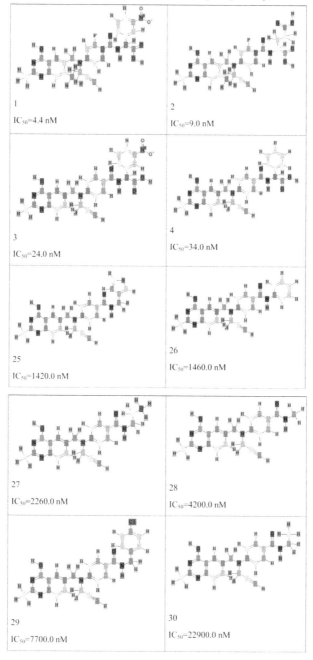

TABLE 21.7   Cyclic Properties Characteristic for Highly, Medium- and Low Effective Antitumor Compounds

Properties characteristic forhighly effective compounds (r≥ 0,1 inM1)		
208	216	228
342	370	403
507	508	509
510	513	514
515	516	519
520	521	522
523	524	525
526	527	528

529	530	531
532	534	535

Properties characteristic for medium effective compounds (r≥ 0,1 inM2)

102	121	131
187	243	343
511	538	544
545	546	561
562	563	

Fragments of low effective antitumor compounds(r≤ -0,1 inM2)

160	200	238
290	298	480

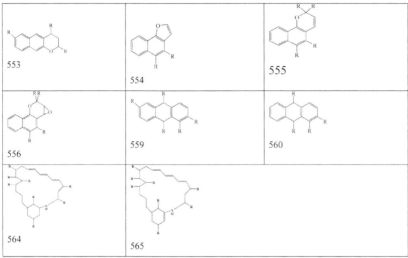

*Figures are used for denoting fragment codes in calculation

t is shown that the initial amine group, oksofosfin group (P=O) and 2,4-disubstituted 2,5-dihydrofuran (cycle 536, Table 21.7) are equally characteristic for medium and highly effective antitumor compounds. In the class of low effective compounds such fragments as a carbonyl group, a silicon atom, 1,2,3,5-tetrasubstituted benzene (cycle 238, Table 21.6), 1,2,3,4-tetrasubstituted pyrrolidine (cycle 200, Table 21.7) are present-ed. For medium effective compounds such properties as a nitro group, a bromine atom, 1,3-disubstituted benzene (cycle 187, Table 21.6) and 1,2,3,5-tetrasubstituted benzene (cycle 243, Table 21.7) are most frequent.

It is established that not only the nature of the fragments contained but also methods of their combination with the neighboring properties influence the effectiveness of the antitumor activity of compounds. The consecutive combination of the carbonyl group with ethylene fragments is typical for low effective antitumor compounds (M2) whereas the combi-nation of the group with 5,6-disubstituted-5,6,7,8-tetrahydro [1,3] dioxol [4,5,9] isoquinoline (cycle 546, Table 21.7) is mainly met in the class of compounds with medium antitumor activity. The same holds true for the combination of two methine groups with the oxygen fragment and mono-substituted phenyl (cycle 102, Table 21.7) and 1,4-disubstituted phenyl (cycle 143, Table 21.6), Fig. 21.1.

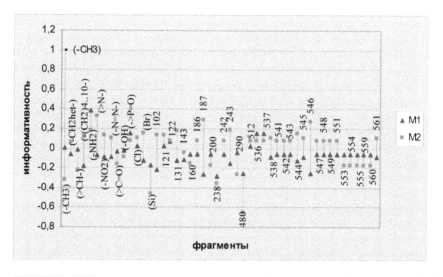

FIGURE 21.1   Acyclic(A) and cyclic fragments (Б), characteristic for highly , medium and low effective antitumor compounds (structural formulas of cyclic fragments are given in Table 21.6).

GUSAR is based on the approach of Quantitative Neighborhoods of Atoms (QNA) and Multilevel Neighborhoods of Atoms (MNA) descriptors as well as on Self-Consistent Regression algorithm. In GUSAR, QNA and MNA descriptors are used to build the consensus model. The obvious limitation of GUSAR is that it never provides the QSAR model as MLR in interpretable from nor any knowledge about the descriptors that are used to build the consensus model.

Statistical Characteristics of Consensus model based on QNA and MNA descriptors from 20 models: N=30, R^2=0.926, F=34.093, SD=0.247, Q^2=0.884, V=6.

Where N is total number of molecules used, R is correlation coefficient, F – is value of Fischer's parameter, SD is standard deviation, the cross-validated R^2 is V is number of variables used in the model building.

The output of GUSAR is in the form of a diagram in which the atoms are colored according to their contribution towards biological activity along with various statistical characteristics used to arrive at the consensus model. Explanation of the colors is as following: "Green" means that the impact of the atom approximately corresponds to the predicted activity value for a whole molecule. "Blue" means that the particular atom may decrease the activity. «Red» means that the particular atom may increase the activity.

To have a better idea, compounds with numbers 1, 2, 3, 4, 25, 26, 27, 28, 29, 30 were used for analysis purpose as representative examples of the same series having higher binding and less binding with enzyme (Table 21.6). The analysis of the diagrams in this table shows that the electron donating and withdrawing groups in the core of these thymidylate synthetase inhibitors on the their inhibitory effectivity are ambiguous.

From Table 21.6, it is evident that contribution of nitrogen atoms in nitrogroup, chlor atom and oxygen atoms of carbonyl fragments to binding with protein are negligible. But fluor atoms and have reverse effect. The sulfur atom in the thiophene ring and carbon atom in –CN group contribute positively. Monosubstituent and 1,4-disubstituend phenyl rings do not play critical role in inhibition activity. Appearance of few "blue" labeled atoms in quinazoline and thiophene rings indicates that these cyclic fragments do not play important role in binding with thymidylate synthetase. The –OH group attached to carbon number 4 in quinazoline ring and –OH-group of carboxylic groups play negative role. Methyl groups do not play negative role in activity.

GUSAR shows good performance and has ability to provide some insight into relative importance of the individual atoms involved in determining the biological activity or with receptor, Fig. 21.2.

The results obtained may serve the basis for the directed synthesis of potential anticancer compounds.

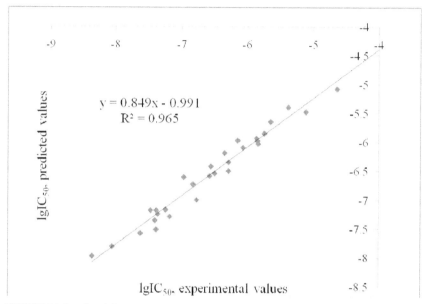

FIGURE 21.2   Graph between observed and predicted values for biologic activity.

21.4   CONCLUSIONS

1. There are revealed the structural properties characteristic for effective and highly effective antitumor compounds.
2. There are built two models of prognosticating interval levels of antitumor activity of sulfur-, nitrogen-, silicon- and oxygen-containing heterocyclic compounds in relation to tumor cells of HepG2, HeLa, HeLa100–6 and DU–145 with the level of valid prediction of 80% by two methods of the pattern recognition theory. The discovered principles of combining structural descriptors allow to carry out a virtual screening of different classes of biologically active substances for the presence of tumor activity in relation to hepatomas, malignant tumors of cervix and prostate.
3. Using QSAR-methodology on basis GUSAR program statistically sound and thoroughly validated consensus model for thymidylate synthetase inhibitors, that are quinazoline derivatives, is obtained. The six parametric model has following statistical characteristics $N=30$, $R^2=0.926$, $F=34.093$, $SD=0.247$, $Q^2=0.884$, $V=6$. It was

found that the effect of electron donating and withdrawing substituents in the core of the quinazoline ring on the their inhibitory effectivity of thymidylate synthetase are ambiguous.

KEYWORDS

- **Antitumor Activity**
- **Descriptors**
- **Pattern Recognition Theory**
- **Thymidylate Synthetase Inhibitors**
- **Virtual Screening**

REFERENCES

1. Son, J. K., Jung, S. J., Jung, J. H., Fang, Z., Lee, C. S., Seo, C. S., Moon, D. C., Min, B. S., Kim, M. R. & Woo, M. H. (2008). Anticancer Constituents from the Roots of Rubiacordifolia L., Chem. Pharm. Bull 56(2), 213–216.
2. Kuwano, M., Ikezaki, K., Mamizuka, K., Komiyama, S., Seto, H., Otake, N., Sugita, M., Yamaguchi, T., Kishiye, T., Fukawa, H. & Sasaki, T. (1983). Potentiation of mitomycin C, 6-mercaptopurine, bleomycin, cis-Diamminedichloroplatinum and 5-fluorouracil by Mycotrienins and Mycotrienols. Gann, 74, 759–766.
3. Takara, K., Sakaeda, T., Yagami, T., Kobayashi, H., Ohmoto, N., Horinouchi, M., Nishiguchi, K. & Okumura, K. (2002). Cytotoxic Effects of 27 anticancer drugs in Hela and MDR1-overexpressing derivative cell lines. Biol. Pharm. Bull. 25(6), 771–778.
4. Kugawa, F., Ueno, A., Kawasaki, M. & Aoki, M. (2004). Evaluation of Cell Death Caused by CDF (Cyclophosphamide, Doxorubicin, 5-Fluorouracil) Multi-drug administration in the human breast cancer cell line MCF–7, Biol. Pharm. Bull. 27(3), 392–398.
5. Saturnino, C., Buonerba, M., Boatto, G., Pascale, M., Moltedo, O., Napoli, L. D., Montesarchio, D., Lancelot, J. C. & Caprariis, P. D. (2005). Synthesis and preliminary biological evaluation of a new pyridocarbazole derivative covalently linked to a thymidine nucleoside as a potential targeted antitumoralagent. I. Chem. Pharm. Bull. 51(8), 971–974.
6. Jiang, M. M., Dai, Y., Gao, H., Zhang, X., Wang, G. H., He, J. Y., Hu, Q. Y., Zeng, J. Z., Zhang, X. K., & Yao, X. S. (2008). Cardenolides Antiaristoxicaria as Potent Selective Nur 77 Modulators. Chem. Pharm. Bull. 56(7), 1005–1008.

7. Santhekadur, P. K., Rajasekaran, D., Siddiq, A., Gredler, R., Chen, D., Schaus, S. E., Hansen U., Fisher, P. B., & Sarkar, D. (2012). The transcription factor LSF: a novel oncogene for hepatocellular carcinoma. Am. J. Cancer Res. 2(3), 269–285.

8. Tyurina, L. A., Tyurina, O. V., & Kolbin, A. M. (2007). Methods and results of projecting and prognosticating biologically active substances. Gilem, Ufa.

9. Koshiura, R., (Hirata), Kagotani, Y., & Ujiie, T. (1962). Experimental anticancer studies. Preparation and anticancer activity of 4-amino-6-hexylresorcinol on Ehrlich Carcinoma in mice. Biochem. J. 10(6), 528–532.

10. Liao, H. L. & Hu, M. K. (2004). Synthesis and anticancer activities of 5,6,7-trimethylbaicalein derivatives. Chem. Pharm. Bull. 52(10), 1162–1165.

11. Jiang, C., You, Q., Liu, F., Wu, W., Guo, Q., Chern, J., Yang, L. & Chen, M. (2009). Desing, synthesis and evaluation of tetra hydro isoquinolines as new kinesin spindle protein inhibitors. Chem. Pharm. Bull. 57(6), 567– 571.

12. Kim, H. S., Kim, Y. H., Yoo, O. J., & Lee, J. J. (1996). Aclacinomycin X, a novel anthracycline antibiotic produced by Streptomyces galilaeus ATCC 31133. Biosci. Biotech. Biochem. 60(5), 906–908.

13. Singh, P., Faridi, U., Srivastava, S., Kumar, J. K., Darokar, M. P., Luqman, S., Shanker, K., Chanotiya, C. S., Gupta, A., Gupta, M. M., & Negi, A. S. (2010). Design and synthesis of C-ring lactone- and lactam- based podophyllotoxin analogues as anticancer agents. Chem. Pharm. Bull. 58(2), 242–246.

14. Hayakawa, I., Shioya, R., Agatsuma, T., & Sugano, Y. (2005). Synthesis and evaluation of 3-methyl-4-oxo-6-phenyl-4,5,6,7-tetrahydrobenzofuran-2-carboxylic acid enhyl ester derivatives as potent antitumor agents. Chem. Pharm. Bull. 53(6), 638–640.

15. Sugimoto, K. & Ohki, S. (1964). Potential anticancer agents. Synthesis of aromatic nitrogen mustards containing azo group. Biochem J. 12(11), 1375–1378.

16. Filimonov, D. A., Zakharov, A. V., Lagunin, A. A., & Poroikov, V. V. (2009). QNA based 'Star Track' QSAR approach. SAR and QSAR Environ. Res. 20(7–8), 679–709.

17. Marsham, P. R.; Jackman, A. L., Barker, A. J., Boyle, F. T., Pegg, S. J., Wardleworth, J. M., Kimbell R., O'Connor, B. M., Calvert, A. H., & Hughes, L. R. (1995). Quinazoline antifolate thymidylate synthase inhibitors: replacement of glutamic acid in the C2-methyl series. J Med Chem. 38, 994–1004.

CHAPTER 22

VIRTUAL SCREENING OF NATURAL AND SYNTHETIC 5-LIPOXYGENASE INHIBITORS

A. YA. GERCHIKOV, V. R. KHAIRULLINA, I. A. TAIPOV, and
F. S. ZARYDIY

CONTENTS

ABSTRACT

Structural features characteristic of high-, medium-, and low-efficiency inhibitors of 5-lipoxygenase (5-LOX) catalytic activity were established. The degree of their influence on the inhibition efficiency was assessed. Two models (Ml and M2) were constructed for the prediction and recognition of the efficiency intervals of 5-LOX inhibitors. The confidence level of activity prediction for Ml and M2 reached 90 and 70%. respectively. The revealed structural features could be used for constructing highly selective inhibitors of 5-LOX catalytic activity.

22.1   INTRODUCTION

The enzyme 5-lipoxygenase is expressed mainly in leukocytes [1] and catalyzes the metabolism of arachidonic acid to form leukotrienes (LTs). The potent biological effects of LTs includes leukocyte aggregation, smooth muscle contraction and vascular permeability which mimic the biological changes associated with the patho-physiology of inflammatory disorders such as asthma, rheumatoid arthritis, inflammatory bowel disease and psoriasis [1, 2].

In addition, the 5-LOX pathway has also been associated with atherosclerosis [2], osteoporosis [1, 2] and certain types of cancer like neuroblastoma [1] and prostate cancer [1]. In this respect, the search for biologically active compounds that are capable of suppressing the activity of 5-LOX represents a promising direction in pharmacology and biochemistry for preventing and treating these pathological states.

Synthetic analogs of leukotrienes in addition to prostaglandins, which arc formed along the cyclooxygenase pathway of arachidonic acid oxidation, arc known to be natural inhibitors of leukotriene biosynthesis [2, 3]. It was reported that flavonoids and other phenolic compounds that are capable of forming stable chelate complexes with Fe^{2+}, which is found at the 5-LOX active center, exhibit a pronounced inhibiting effect for 5-LOX [2–4]. Non-steroidal anti-inflammatory drugs (NSAIDs) and plant extracts exhibit insignificant inhibiting activity on polyunsaturated fatty acid (PUFA) oxidation by 5-LOX [2–5]. Thus, a significant volume of information on this problem has accumulated by now in the domestic and foreign literature. However, the structure—activity relationship of the various classes of 5-LOX inhibitors has not been systematically analyzed. All results from searches for efficient inhibitors of leukotriene biosynthesis catalyzed by 5-LOX are disparate in nature. Therefore, the structure-activity relationship in a series of natural and synthetic 5-LOX inhibitors was studied in the present work in order to discover the structures of novel compounds with high inhibiting activity.

22.2 MATERIALS AND METHODS

The computer system SARD-21 (Structure Activity Relationship & Design) that implements the basic principles of sample recognition theory was used to investigate the structure-activity relationship [6]. Two models for predicting and recognizing interval levels of inhibiting activity of potential different classes of N-. O-. and S-containing 5-LOX inhibitors were constructed in terms of the basic procedures of the SARD-21 system.

Model Ml was designed to recognize compounds with moderate inhibiting activity for 5-LOX. Model M2 was directed al revealing high-efficiency 5-LOX inhibitors. Training sets for both models were constructed according to a dichotomous procedure. Classification was carried out based on results from a comparison of literature data for the inhibiting activity of various biologically active substances for 5-LOX. The parameter for 50% inhibition of isomeric forms of 5-LOX (IC_{50}) isolated from basophils of rat leukemia cells was used as the criterion for assigning initial compounds to the high- or low-efficiency inhibitor class. Series A for model M1 contained 51 efficient 5-LOX inhibitors ($IC_{50} \leq 5.5$ M). Series B included 47 compounds with low efficiency inhibiting activity ($IC_{50} \geq 7.0$ M) [7–16].

Model M2 included 40 high-efficiency 5-LOX inhibitors ($IC_{50} \leq 2.5$ M) and 46 medium- and low-efficiency compounds with $IC_{50} > 2.5$ M (Class B). Table 22.1 presents typical structures of compounds in the training sets of models M1 and M2 in addition to the IC_{50} values corresponding to them.

TABLE 22.1 Typical Structures of Compounds in the Training Set for Constructing Models M1 and M2

Model for predicting and recognizing medium-efficiency and low-efficiency 5-LOX inhibitors (M1)			
Class of active compounds (medium-efficiency 5-LOX inhibitors)			
$IC_{50} = 0.5\text{-}1\mu M$	$IC_{50} = 1\ \mu M$	$IC_{50} = 0.7\ \mu M$	$IC_{50} = 0.35\ \mu M$
$IC_{50} = 0.4\ \mu M$	$IC_{50} = 0.1\ \mu M$	$IC_{50} = 0.64\ \mu M$	$IC_{50} = 0.36\ \mu M$
Class of inactive compounds (low-efficiency 5-LOX inhibitors)			
$IC_{50} = 7.13\ \mu M$	$IC_{50} = 15\ \mu M$	$IC_{50} = 10.4\ \mu M$	$IC_{50} = 43.1\ \mu M$
$IC_{50} = 17.3\ \mu M$	$IC_{50} = 16.8\ \mu M$	$IC_{50} = 16.6\ \mu M$	$IC_{50} > 100\ \mu M$
Model for predicting and recognizing high-efficiency and medium-efficiency 5-LOX inhibitors (M1)			
Class of active compounds (high-efficiency 5-LOX inhibitors)			

IC$_{50}$=0.06±0.017 µM	IC$_{50}$=0.08 µM	IC$_{50}$=0.13 µM	IC$_{50}$=0.34±0.11 µM
IC$_{50}$=0.81 µM	IC$_{50}$=0.2 µM	IC$_{50}$=0.03 µM	IC$_{50}$=0.1 µM
Class of inactive compounds (medium-efficiency 5-LOX inhibitors)			
IC$_{50}$=2.5±0,09 µM	IC$_{50}$=4.8 µM	IC$_{50}$=2.9 µM	IC$_{50}$=9.0 µM
IC$_{50}$>100 µM	IC$_{50}$>100 µM	IC$_{50}$>100 µM	IC$_{50}$>100 µM

The boundary criteria between high- and medium-efficiency classes were the numerical IC$_{50}$ values for a natural compound in the M1 training series, curcumin (IC$_{50}$ = 2.7 M, Tables 22.2 and 22.3), and the active ingredient of the preparation zileuton, N-(1-benzothien-2-ylethyl)-N-hydroxyurea (IC$_{50}$ = 0.9 M, Table 22.3), which are reference selective inhibitors of 5-LOX enzyme activity [11].

TABLE 22.2 Decisive Set of Descriptors Model M1

No.	Descriptors content	r
1	{(-OH) – (>C=C<)} I {(>N-) – (-OH)} I {(-CH$_2$het-) – (>C<)}	0.625
2	{(-CH$_3$) – (-O-)} I {(-CC-) – (p-substituted benzene)} I {(-CH$_2$het-) – (>C<)}	0.463
3	{(>N-) – (-OH)} I {(-CH$_2$het-) – (>C=O)} I {(>C=O) – (-C°C-)}	0.462
4	(-OH) I (-O-) I (-CH$_2$het-)	0.446
5	{(-O-) – (>C=C<)} I {(-CC-) – (p-substituted benzene)} I {(-CH$_2$het-) – (>C<)}	0.441

TABLE 22.2   *(Continued)*

No.	Descriptors content	r
6	{(>C=O) – (-OH)} I {(>N-) – (>C=C<)} I {(-O-) – (>C=O)}	–0.620
7	{(-N=C<) – (>C=C<)} I {(>N-) – (>C=C<)} I {(>CH-) – (>C<)}	–0.552
8	(-N=C<) & (1,2,4-trisubstituted-1,4-dihydroquinoline) & (monosubstituted benzene)	–0.497
9	(1,2,4-trisubstituted-1,4-dihydroquinoline) & (6-substituted quinoline) & (monosubstituted benzene)	–0.417
10	(-CH$_2$-) & (monosubstituted benzene) & (-(CH$_2$)4…10-)	–0.392
11	(monosubstituted benzene) & (>C<) & (Cl)	–0.359

Here, & – conjunction descriptor (logical "and");
! – disjunction descriptor (logical "or");
– strict disjunction descriptor (logical "or not").

TABLE 22.3   Decisive Set of Descriptors Model M2

No.	Descriptors content	r
1	{(-S-) – (>C=C<)} I {(-CH$_2$het-) – (-O-)} I {(>CH-) – (1,4-disubstituted benzene)}	0.695
2	(2-substituted 1-benzothiophene) & (4-substituted morpholine) &(2,11-disubstituted 6,11-dihydroand dibenzo[*b,e*] oxepine)	0.604
3	{(>CH-) – (2-substituted 1-benzothiophene)} I {(-CH$_2$-) – (-CH$_2$het-)} I {(-CH$_2$het-) – (2,11-disubstituted 6,11-dihydro- and dibenzo[*b,e*] oxepine)}	0.604
4	(4-substituted morpholine) & (-S-)& (2,11-disubstituted 6,11-di- hydro- and dibenzo[*b,e*]oxepine)	0.588
5	(-S-) & (2,11-disubstituted 6,11-dihydro- and dibenzo[*b,e*]ox- epine) & (1,2-disubstituted 1-*p*-imidazole)	0.517
6	{(-CH$_2$-) – (-CH$_2$het-)} I {(-CH$_2$het-) – (2,11-disubstituted 6,11-dihydro- and dibenzo[*b,e*]oxepine)} I {(>CH-) -(1,4-disubstituted benzene)}	0.512
7	{(>C=C<) – (4-substituted morpholine)} I {(-CH$_2$het-) – (2,11-disubstituted 6,11-dihydroand dibenzo[*b,e*] oxepine)} I {(>CH-) – (1,4-disubstituted benzene)}	0.496

TABLE 22.3   *(Continued)*

No.	Descriptors content	r
8	{(-CH$_2$het-) – (-O-)}I {(-NH-) – (>C=O)}I	0.487
	{(>CH-) – (>N-)}	
9	{(>C=C<) – (>C=C<) – (>C=C<)}	0.480
10	{(>CH-) – (>C=O)} I {(-CH$_3$) – (>C=C<)} I	–0.486
	{(-CH$_2$het-) – (>CH-)}	
11	{(-NH-) – (>C=C<)} I {(-CH$_3$) – (>C=C<)} I	–0.469
	{(-CH$_2$het-) – (>CH-)}	
12	{(>C=O) – (-OH)} I {(-CH$_3$) – (>C=C<)} I	–0.469
	{(-CH$_2$het-) – (>CH-)}	
13	{(>CH-) – (-(CH$_2$)4.10-)} I {(-CH$_3$) – (>C=C<)} I	–0.469
	{(>CH-) – (>CH-)}	
14	(209) & (2-substituted furan) & (SO$_2$)	–0.468
15	(-NH$_2$) & (Cl) & (1-substituted benzene)	–0.447
16	(2-substituted furan) & (6-substituted quinoline) & (SO$_2$)	–0.417
17	(6-substituted quinoline) & (-NH$_2$) & (1-substituted benzene)	–0.389
18	(2,4,6-trisubstituted pyrimidine) & (-NH) &	–0.389
	(1-substituted benzene)	

Here, & – conjunction descriptor (logical "and");

! – disjunction descriptor (logical "or");

– strict disjunction descriptor (logical "or not").

Also, the majority of researchers consider compounds with IC$_{50}$ > 7 – 10 M to be medium-efficiency 5-LOX inhibitors [12–14]. The deciding set of descriptor (DSD) of the two models was tested on structures of two examination sets containing 25 and 22 compounds for models M1 and M2, respectively, with known inhibiting activity for 5-LOX. Data on their inhibiting activity were obtained under analogous experimental conditions by binding to rat blood cells [11–14]. The basic principles implemented in SARD-21 did not enable the analyzed structures to be examined as a sequence of chemically bonded atoms. Rather, each structure was represented as a set of fragment descriptors (FD). As a result, the stereochemistry was not considered in the analysis of the structural information.

The following types of FD were examined during the structure analysis: 1) initial fragments including elements of cyclic systems and the cyclic systems themselves; 2) substructure descriptors from several chemically bonded initial fragments; and 3) logical combinations (conjunctions, disjunctions, strict disjunctions) generated based on the first and second type of descriptors [17]. This method for representing the structures was used in the analysis of compounds of the training and examination sets and of predicted structures. Such an approach to recognizing structures did not model prediction of the target specificity property for any actual class of organic compounds. Instead, it enabled the obtained models to be considered sufficiently universal for various classes of structurally heterogeneous biologically active compounds.

The nature of the FD effect on the efficiency of 5-LOX inhibition was estimated from the information-value coefficient r (correlations of qualitative descriptors) $(-1 < r < 1)$. According to this, the greater the positive value of the information value was, the greater was the probability of this descriptor affecting the manifestation of the target property (positive and negative descriptors, respectively, "+" and "–") [6]. The information-value coefficient of descriptors r was calculated in terms of the corresponding SARD-21 procedures using the formula:

$$r = (n_1 n_4 - n_2 n_3)/(\sqrt{N_1 N_2 (n_1 + n_3)(n_2 + n_4)}),$$

where n_1 and n_2 are the number of structures in the inactive compound class containing and not containing this fragment; n_3 and n_4, the number of structures in the active compound class containing and not containing this fragment; N_1 and N_2, the number of structures in the active and inactive compound classes, respectively [6]. Thus, the sensitivity of the models was determined by the structural characteristics of the training sets, more accurately, by features of the structure and number of active and inactive structures contained in them.

Recognition and prediction models for the studied type of activity were formulated as a result of combining classification rules and the deciding set of structural parameters as logical equations of the type $C = F(S)$, where C is the property (activity); F, recognition rules (recognition algorithm of samples according to which the studied compounds were classified, geometric or voting); and S, the DSS. The geometric approach was based on a determination of the distances to certain standards in multidimensional DSD space using a Euclidean metric. The voting method was based on a comparison of the number of positive and negative descriptors of the deciding set that described each analyzed structure. The efficiency of the models for the studied types of activity was determined from results of testing compounds of the examination set and structures of the initial series using the two methods of sample recognition theory, the geometric approach and the voting method [6].

22.3   RESULTS AND DISCUSSION

Fragment descriptors and their logical combinations that were potentially responsible for the manifestation of the studied type of activity were included in the DSD of each of the constructed models (Tables 22.2 and 22.3) through automated selection in terms of the used algorithm. Descriptors with positive information-value coefficients in model M1 were characteristic of medium-efficiency 5-LOX inhibitors; with negative values, of low-efficiency 5-LOX inhibitors. According to DSD data for model M2, a positive information-value coefficient corresponded to descriptors of high-efficiency 5-LOX inhibitors whereas descriptors with a negative information-value coefficient were characteristic of medium-efficiency 5-LOX inhibitors.

Table 22.4 presents results from recognition of training and examination sets using DSD formulated for models M1 and M2. These data indicated that models M1 and M2 constructed by us had high recognizing capabilities and; therefore, could be used to predict interval levels of inhibiting activity of novel compounds as 5-LOX inhibitors.

TABLE 22.4   Results of Recognition (%) of Training and Examination Sets Using Deciding Sets of descriptor (DSD) Formed for Models M1 and M2

Recognition method	DSD for model M1				DSD for model M2			
	Series A	Series B	All set	Exam-ination set	Series A	Series B	All set	Exami-nation set
Geometric	78%	79%	79%	77%	74%	77%	76%	82%
Voting	78%	77%	77%	84%	80%	91%	85%	72%

Table 22.5 and Table 22.6 present the fragment of the training set with results of structure recognition.

TABLE 22.5 Results of Testing DSD of Model M1 on Examination Set

1		A	A	8.5
2		A	A	300
3		A	A	3.9
4		A	A	10
5		A	A	0.5-0.9
6		A	A	22
7		A	A	320
8		A	A	<30
9		A	A	4.2
10		A	A	>3
11		A	A	3.6-8.8
12		A	A	>100

13		A	A	3,2
14		A	A	1.37-8.9
15		A	A	2-10
16		A	A	6.9
17		A	A	3
18		A	A	150-200
19		B	A	5
20		B	A	0.5-1
21		B	A	60
22		B	B	7.7-11.4
23		B	B	29
24		B	B	1.5-15
25		B	B	>3

TABLE 22.6   Results of Testing DSD of Model M2 on Examination Set

№	Structural formula	Recognition by voting method	Recognition by geometric method	Literature IC50, µM
1		A	A	0,06±0,017
2		A	A	0,34±0,11
3		A	A	0,195
4		A	A	0,54
5		A	A	6,8
6		A	A	5,5
7		A	A	3,8
8		A	A	4,9
9		A	A	5,5

No.	Structure			
10		A	A	1,19
11		B	A	3,8
12		A	A	6,6
13		A	A	5,6
14		A	A	3,5
15		A	A	2,1
16		A	A	2,8
17		B	A	4,9
18		B	B	4,8
19		B	B	6
20		B	B	3,9
21		B	B	8,4
22		A	B	3,9

Joint analysis of FD of both models revealed fragments characteristic of high- and medium-efficiency 5-LOX inhibitors with information-value coefficients $r - 0.1$ that were found using models M1 and M2 in addition to descriptors of low-efficiency 5-LOX inhibitors with information-value coefficient $r-0.1$ as estimated from model M1. Their effects were analyzed considering the assignment to various functional groups. Table 22.7 presents cyclic descriptors characteristic of medium- and high-efficiency 5-LOX inhibitors.

TABLE 22.7   Cyclic Descriptors Characteristic of High- and Medium-Efficiency 5-LOX Inhibitors

\n112	\n116	\n122	\n550
\n552	\n554	\n551	

Fragments of low-efficiency 5-LOX inhibition			
\n582	\n108	\n275	\n102
\n584	\n586	\n209	\n162
\n136	\n108	\n121	\n553
\n340	\n551		

* Fragment codes in the calculations are denoted by numbers.

* Fragment codes in the calculations are denoted by numbers.

Cyclic fragments that were encountered with an equal probability level in the medium- and low-efficiency classes, the numerical information-value coefficient of which was in the range $-0.1 < r < 0.1$, appear in this same table.

It was established that descriptors such as hydroxyl and methyl groups and O and Br atoms were most characteristic of functional groups for medium-efficiency 5-LOX inhibitors. Simple acyclic descriptors such as a

tertiary N atom and S, F, and Br atoms were characteristic of high-efficiency 5-LOX inhibitors (Fig. 22.1).

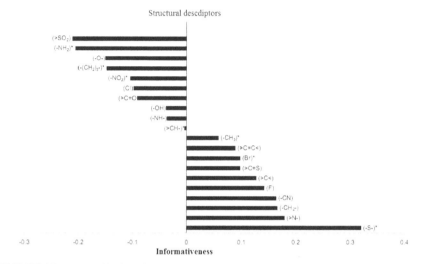

FIGURE 22.1   Acyclic descriptors characteristic of high- and medium-efficiency 5-LOX
inhibitors.
*Denotes fragments characteristic of high-efficiency 5-LOX inhibitors.

It was shown that not only the nature of the fragment but also the bonding mode to neighboring descriptors had an important influence on the
efficiency of inhibiting 5-LOX catalytic activity. Thus, sequential combination of amines with two ethylene fragments was characteristic of
medium-efficiency 5-LOX inhibitors (Fig. 22.2) whereas the combination
of this same group with an ethylene and O had a negative effect on the
manifestation of inhibiting activity.

It should be noted that combinations of 1,3,5-trisubstituted benzene
(ring 238 in Fig. 22.2) with N-, O-, and S-containing cyclic and acyclic
descriptors were characteristic of medium-efficiency 5-LOX inhibitors.
An analysis of the effect of the surroundings on the efficiency of 5-LOX
inhibition in terms of model M2 showed that sequential combination of a
tertiary N atom with ethylene and azamethine groups was encountered pri-
marily in the high-efficiency class of compounds. However, its combination with ethylene and carbonyl groups was characteristic of medium- and
low-efficiency compounds (Fig. 22.3).

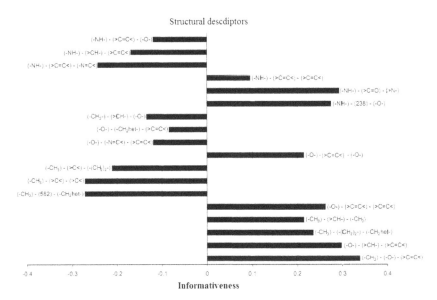

FIGURE 22.2 Effect of N-, O-, and S-containing descriptors on efficiency of inhibiting activity for 5-LOX according to M1.

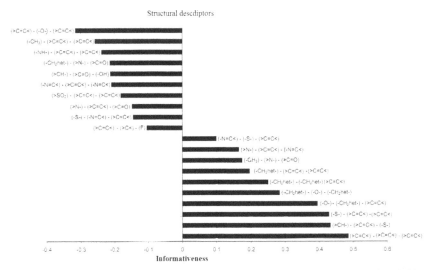

FIGURE 22.3 Effect of N-, O-, and S-containing descriptors on efficiency of inhibiting activity for 5-LOX according to M2.

The results enabled virtual screening of various classes of N-, S-, and O-containing biologically active compounds to be carried out in order to find potentially selective 5-LOX inhibitors. It was obvious that if the structural descriptors identified by us were missing in the molecules intended for prognosis, then these molecules would not fall in the range of applicability of the model.

The discovered features in the structures of high-efficiency inhibitors of 5-LOX catalytic activity could provide a basis for targeted synthesis of potential inhibitors of this enzyme.

KEYWORDS

- **5-lipoxygenase**
- **inhibitors of catalytic activity**
- **structure activity**

REFERENCES

1. Werz, O., & Steinhilber, D. (2005). Biochemical Pharmacology, 70 327–333.
2. Boyington, J. C., Gaffney, B. J., & Amzel, L. M. (1985). Pharmacology, 38, 199–205.
3. Pontiki, E., & Hadjipavlou-Litina, D. (2005). Curr. Enzyme Inhib. 1, 309–327.
4. Webb, E. C. (1992). Enzyme Nomenclature, Academic Press, San Diego.
5. Jones, G. D., Russell, L., Darley-Usmar, V. M., Stone, D., & Wilson, M. T. (1996). Biochemistry, 35, 7197–7203.
6. Tapiero, H., Townsend, D., & Tew, K. (2004). Biomed Pharmacother, 58, 183–193.
7. Tyurina, L. A., Tyurina, O. V., & Kolbin, A. M. (2007). Methods and Results of the Design and Prediction of Biologically Active Compounds [in Russian], Gilem, Ufa.
8. Block, E., Iyer, R., Grisoni, S., et al. (1988). J. Am. Chem. Soc., 110, 7813–7827.
9. Prigge, S. T., Boyington, J. S., Faig, M., & Doctor, K. S. (1997). Biochimie 79, 629–636.
10. Vasques-Martinez, Y., Ohri, R. V., Kenyon, V., & Holman, T. R. (2007). Bioorg. Med. Chem., 15, 7408–7425.
11. Camargo, A. B., & Marchevsky, E. (2007). J. Agric. Food Chem., 55, 3096–3103.
12. Werz, O. (2007). Planta Med. 73, 1331–1357.
13. Rao, P. N. P, Chen, Q.-H., & Knaus, E. E. (2006). Med. J., Chem., 49, 1668–1683.
14. Singh, U. P., Prithiviraj, B., Sarma, B. K., et al. (2001). Indian J. Exp. Biol., 39, 310–322.
15. U. P. Kelavkar, W. Glasgow, S. J. Olson, and B. A. Foster, Neoplasia, 6, 821–830 (2004).

16. Hsieh, R. J., German, J. B., & Kinsella, J. E., (1988). Lipids, 23, 322–326, Werz, O., Szellas, B., & Steinhilber, D. (2000). Eur. J. Biochem., 267, 1263–1269.
17. Denisov, E. T., & Afanasev, I. B. (2005). Oxidation and Antioxidants in Organic Chemistry and Biology, Talyor & Francis, Boca Raton.

CHAPTER 23

KINETICS OF LIQUID-PHASE OXIDATION OF POLYVINYL ALCOHOL

YU. S. ZIMIN, I. M. BORISOV, and A. G. MUSTAFIN

CONTENTS

ABSTRACT

It is shown that polyvinyl alcohol oxidation at influence of ozone-oxygen mixture and hydrogen peroxide in an aqueous medium is carried out by the radical mechanism. This process is accompanied by oxidative destruction and functionalization of polymer macromolecules and oligomers. The scheme of liquid-phase oxidation of polyvinyl alcohol is suggested

23.1 INTRODUCTION

Polyvinyl alcohol (PVA) finds its application in various activity spheres including medicine [1]. PVA water solubility, nontoxicity and full indifference to living organism's tissues is promoted by its multiform usage in medical practice. Scientific interest to polyvinyl alcohol is caused by using its low molecular weight products subjected to additional functionalization as polymer substrates of prolonged effect drugs.

Due to the abovementioned the obtaining of PVA modified oligomers of small molecular weight is quite urgent. In modifying polyvinyl alcohol by metal compounds [2–10] and some acids [11, 12] the received reaction products are to be additionally purified from remnants of these reagents that raises the cost price of the final product. The ozone-oxygen mixture and hydrogen peroxide are deprived of these drawbacks and allow to obtain modified oligomers suitable for further use without any additional cleaning.

In the given work the data of kinetic laws of oxidation, oxidative destruction and oxidative functionalization of polyvinyl alcohol under the influence of the ozone-oxygen mixture (reaction system "PVA+ O_3 + O_2 + H_2O") and hydrogen peroxide (reaction system "PVA + H_2O_2 + O_2 + H_2O") in aqueous solutions are presented.

23.2 EXPERIMENTAL PART

For experiments polyvinyl alcohol of "Sigma-Aldrich Corporation" (St. Louis, MO, USA) with a molecular weight of 68,000 ([η] = 1.1 dl/g, 25°C water) is used. An ozone-oxygen mixture was received by using an ozonizer of the known construction [13]. Hydrogen peroxide was used of "analytical reagent grade" type and ferrous sulfate (II) and disodium salt of ethylene diaminetetra acetic acid (Trilon B) of "chemically pure" type. Freshly redistilled water served as a solvent.

Polyvinyl alcohol oxidation was carried out in a glass thermostated reactor of a bubbling type with periodical samples taking. The kinetics of the process was monitored by measuring the concentration of peroxyl (iodimetry) and carboxyl (alkalimetry) groups.

Experiments on carbon dioxide accumulation were carried out in circulation systems with a gas chromatography analysis of the gas sphere. Kinematic and intrinsic viscosities of PVA water solutions were measured in a Ubbelohde viscosimeter with the hung level.

Chemiluminescence in the visible spectral region was watched with the use of an installation comprising a light-tight camera, a photomultiplier C–11, a thermostatic box, an optically transparent glass reactor, a mixer and an automatic recorder. The experiments were made in the following way. The aqueous solution with the definite amount of PVA and $FeSO_4$ was put in the reactor and thermostated in it for 15 min. While mixing hydrogen peroxide was added with a syringe. Any changes in chemiluminescence signal intensity were registered on a diagram strip of the recorder.

Mathematical processing of the experimental results was carried out for the 95% confidence interval.

23.3   RESULTS AND DISCUSSION

Heating of PVA water solutions for 70–90°C does not lead to any changes in the solution viscosity and chemical transformations of polymer macromolecules. However, when introducing ozone or hydrogen peroxide (able to initiate the radical process) into PVA water solutions, oxidative and destructive polymer transformations take place.

1. Polyvinyl alcohol oxidation in the reaction systems "PVA + O_3 + O_2 + H_2O" and "PVA + H_2O_2 + O_2 + H_2O" is confirmed by the acid (Fig. 23.1) and carbon dioxide (Fig. 23.2) accumulation. The analysis of the kinetic curves of carboxyl groups accumulation (Fig. 23.1) showed that they are best described by the equation:

$$[\text{-COOH}] = a \times t + b \times t^2], \tag{I}$$

where a and b are effective parameters characterizing minor and basic accumulation channels of COOH-groups respectively. Carbon dioxide is accumulated by the law (Fig. 23.2):

$$[CO_2] = c \times t^2, \tag{II}$$

where c is an effective parameter characterizing dynamics of CO_2 accumulation.

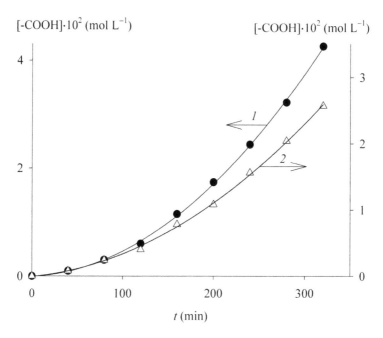

FIGURE 23.1 Kinetic curves of carboxyl groups accumulation in the reaction systems "PVA + O_3 + O_2 + H_2O" (*1*) (90°C, $[PVA]_0$ = 3.5% mass.) and "PVA + H_2O_2 + O_2 + H_2O" (*2*) (90°C, $[PVA]_0$ = 2% mass., $[H_2O_2]_0$ = 0.8 mol/L).

The comparison of the (I) and (II) equations indicates that in the process of PVA oxidative transformations the basic part of carboxyl groups and carbon dioxide are accumulated by the same quadratic law. Basing on the fact it is possible to suppose that the final PVA oxidation products (acids and CO_2) are formed in parallel (non-limiting) stages.

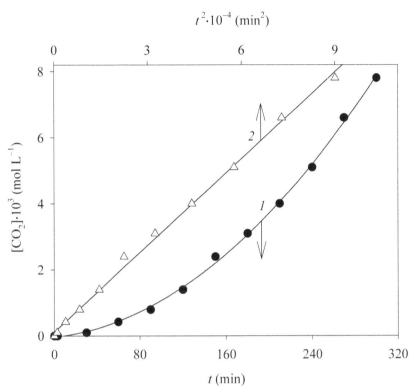

FIGURE 23.2 Kinetic curve of carbon dioxide accumulation in the reaction system "PVA + H_2O_2 + O_2 + H_2O" (1) and its straightening in the equation coordinates (II) (2) (90°C, $[PVA]_0 = 2\%$ mass., $[H_2O_2]_0 = 1$ mol/L).

Carbon dioxide in the reaction system "PVA + H_2O_2 + O_2 + H_2O" is dissolved by the law of the first order reaction:

$$[H_2O_2] = [H_2O_2]_0 \times e^{-kt}, \qquad (III)$$

where k is an effective speed constant of H_2O_2 consumption.

While increasing the initial polyvinyl alcohol concentration in the system "PVA + H_2O_2 + O_2 + H_2O" the effective speed constant of H_2O_2 consumption k changes within the experimental error (Table 23.1). It means that polyvinyl alcohol and products of its oxidation do not influence on the expenditure of hydrogen peroxide. At the same time the parameter values a, b, and c pass through a maximum and change the initial PVA concentration in the reaction systems "PVA + O_3 + O_2 + H_2O" and "PVA +

$H_2O_2 + O_2 + H_2O$" (the dependence of the parameter b from the PVA initial concentration are given in Fig. 23.3). Carboxyl groups and carbon dioxide are formed in the process of oxidative PVA transformations and thus, in increasing $[PVA]_0$ the speed of COOH-groups and CO_2 accumulation grows as well. Decreasing of the parameters a, b, and c in high initial concentrations of the polymer ($[PVA]_0 > 3,5\%$ mass) is probably connected with the increase in viscosity characteristics of the solution leading to diffusive limitations of the oxidative process.

TABLE 23.1 Dependence of the Effective Speed Constant k of H_2O_2 Expenditure from the Initial Concentration of PVA (90°C, $[H_2O_2]_0 = 1.5$ mol/L)

$[PVA]_0$, % mass.	0.5	1	2	3	4	5	7
$k \cdot 10^4$, min^{-1}	3.8 ± 0.2	4.0 ± 0.1	4.2 ± 0.2	4.4 ± 0.2	4.4 ± 0.1	4.4 ± 0.3	4.0 ± 0.4

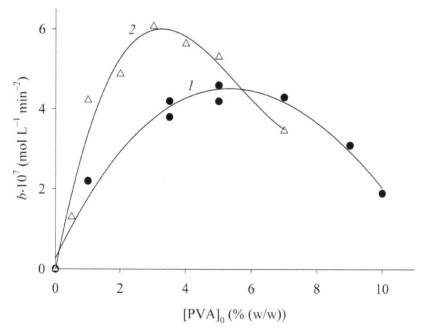

FIGURE 23.3 Dependence of b parameter from the initial PVA concentration in the reaction systems "PVA + O_3 + O_2 + H_2O" (1) (90°C) and "PVA + H_2O_2 + O_2 + H_2O" (2) (90°C, $[H_2O_2]_0 = 1.5$ mol/L).

The parameter values of k, b and c are linearly dependent of the temperature ($70 \div 95°C$) in the Arrhenius coordinates:

$\lg b = (-1.1 \pm 0.5) - (8.8 \pm 0.7)/\theta$ (reaction system "PVA + O_3 + O_2 + H_2O"),

$\lg k = (11 \pm 1) - (22 \pm 2)/\theta$ (reaction system "PVA + H_2O_2 + O_2 + H_2O"),
$\lg b = (10 \pm 3) - (26 \pm 5)/\theta$ (reaction system "PVA + H_2O_2 + O_2 + H_2O"),
$\lg c = (9 \pm 4) - (24 \pm 6)/\theta$ (reaction system "PVA + H_2O_2 + O_2 + H_2O"),

where $q = 2.303\ RT$ kcal/mol. From the above data it may be concluded that the effective parameters b and c characterizing the accumulation speed of carboxyl groups and carbon dioxide in the same reaction system "PVA + H_2O_2 + O_2 + H_2O" obtain merely the same activation parameters. This fact may serve an argument in favor that acids and carbon dioxide are formed in parallel non-limiting stages.

2. Oxidative destruction of PVA macromolecules is supported by kinetics of changes in kinematical and intrinsic viscosities in aqueous polymer solutions (Fig. 23.4). Comparing the data of this figure with the kinetics of carboxyl group (Fig. 23.1) and carbon dioxide (Fig. 23.2) accumulation shows that while viscosity decreases quite rapidly on the initial stage of PVA oxidation, the concentration of COOH-groups and CO_2 considerably increases. On the next stage, it is vice versa: the solution viscosities slightly change and the accumulation speed of the final reaction products rise sharply. From the above data it is concluded that the oxidative destruction prevails on the initial stage of PVA oxidation whereas oxidation of the obtained destruction products dominates on the next stage.

The degree of reducing the kinematical viscosity during the first 40 min depends on the hydrogen peroxide content in the reaction system and it respectively increases whereas $[H_2O_2]_0$ raises (Table 23.2). With increasing temperature and introducing metal salts of variable valence ($FeSO_4$) in the reaction system "PVA + H_2O_2 + O_2 + H_2O" the degree of kinematical viscosity reduction raises as well (Tables 23.3 and 23.4).

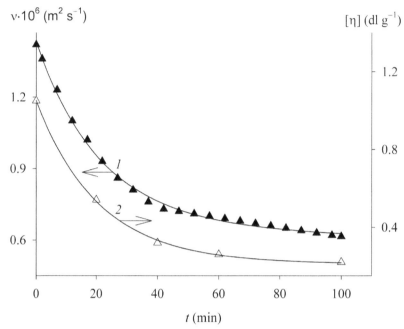

FIGURE 23.4 Dependence of kinematical (*1*) and intrinsic (*2*) viscosities from time in the reaction system "PVA + H_2O_2 + O_2 + H_2O" (80°C, $[PVA]_0$ = 2% mass., $[H_2O_2]_0$ = 1.5 mol/L).

TABLE 23.2 Dependence of Decreasing Degree of Kinematical Viscosity of PVA Aqueous Solutions from H_2O_2 Initial Concentration (80°C, $[ПBC]_0$ = 2% mass.)

$[H_2O_2]_0$, mol/L	0.5	1.0	1.5	2.0
$\dfrac{V_0 - V_t}{V_0} \times 100, \%$ *	44	46	49	52

*V_0 and V_t - kinematical viscosities of PVA solutions in the initial moment and t = 40 min, respectively.

TABLE 23.3 Dependence of decreasing degree of kinematical viscosity of PVA water solutions from the temperature ($[PVA]_0$ = 2% mass., $[H_2O_2]_0$ = 1.5 mol/L)

T, °C	60	70	80	90
$\dfrac{V_0 - V_t}{V_0} \times 100, \%$ *	42	45	49	56

*V_0 and V_t - kinematical viscosities of PVA solutions in the initial moment and t = 40 min, respectively.

TABLE 23.4   Dependence of Decreasing Degree of Kinematical Viscosity of PVA Water Solutions From $FeSO_4$ Additives (90°C, $[PVA]_0$ = 2% mass., $[H_2O_2]_0$ = 1 mol/L)

$[FeSO_4]_0 \cdot 10^4$, mol/L	0	2
$\dfrac{v_0 - v_t}{v_0} \times 100, \%$ *	16	63

*v_0 and v_t - kinematical viscosities of PVA solutions in the initial moment and t = 10 min, respectively.

3. Radical nature of the process. The quadratic time dependences of acid accumulation (the second summand of the equation (I)) and carbon dioxide (equation (II)) in the researched systems definitely point to the fact that the stages of accumulation of the intermediate product [14, 15] must precede the stage of these substances formation. The kinetics of acids and CO_2 accumulation is impossible to explain when considering the polyvinyl alcohol oxidation as a redox reaction. Polyvinyl alcohol here acts a reducing agent and hydrogen peroxide or ozone is an oxidant. If this reaction were considerable, the speed of acid and carbon dioxide accumulation would be maximum at the beginning of the reaction and gradually reduce during the process. According to the Figs. 23.1 and 23.2, the accumulation speed of CO_2 and COOH-groups conversely increases during the reaction. Thus, ozone and hydrogen peroxide obviously serve as initiators in the oxidative transformations of the polymer.

The radical nature of PVA oxidation is supported by the researches carried out by the reaction system "PVA+ H_2O_2 + O_2 + H_2O":

1. Polymer oxidation proceeds only in the presence of metal impurities of variable valence contained in the used bi-distilled water and the source PVA[1]* (Table 23.5). While introducing 1×10^{-3} mol/L of Trilon B (capable of binding metal ions in the complex) in the system of "2% mass. PVA +1.5 mol/L H_2O_2 + H_2O" in 90°C, the speed of hydrogen peroxide expenditure decreases whereas carboxyl groups stop accumulating. Thus, disintegration of hydrogen peroxide is catalyzed by metal impurities of M^{n+} variable valence in H_2O and PVA, leading to formation of active

---

*The quantitative analysis of metal impurities was fulfilled by the method of atomic adsorption spectroscopy in the analytical chemistry department of Research Institute of Life Safety in the Republic of Bashkortostan.

HO˙ radicals and reaction (1) is obviously a stage of initiating the radical oxidation of polyvinyl alcohol. The additives of Trilon B bind M^{n+} ions in a non-active complex resulting in the stop of the oxidative process.

$$H_2O_2 + M^{n+} HO˙ + HO^- + M^{(n+1)+}, \tag{1}$$

TABLE 23.5 Content of Variable Valence Metals in the Components of the Reactionary Systems

Metal	PVA^a	H_2O
Iron	0	0.021
manganese	0	<0.01
copper	0.043	0.081
chromium	0	<0.02

$^a[PVA]_0 = 1\%$ mass.

2. As the initiation speed (see reaction (I)) is proportional to the M^{n+} concentration, the effective parameters characterizing the speed of the oxidative process must increase while introducing metal ions of variable valence into the reaction system "PVA + H_2O_2 + O_2 + H_2O." In introducing a ferrous sulfate (II) into the researched system, the effective speed constant k of hydrogen peroxide expenditure and a, b, and c parameters of acid and CO_2 accumulation increase, respectively (on Fig. 23.5 there are given the dependencies of the effective parameters k and c from $[FeSO_4]_0$).

The initiation speed of the reaction (I) influences the speed of destructive processes that can be seen in the Table 23.2–23.4. The rise in the initial concentrations of H_2O_2, $FeSO_4$ and the temperature really increases the concentration of PVA radical intermediators subjected to destruction in the reaction system. It leads in its turn to lowering the dependence of kinematic viscosity from the mentioned factors (Table 23.2–23.4).

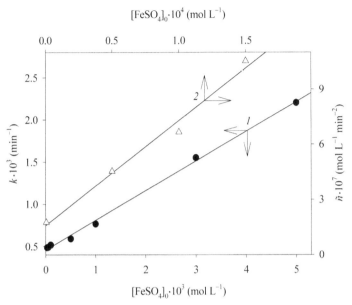

FIGURE 23.5 Dependence of the effective speed constant k of spending H_2O_2 (*1*) and c parameter (*2*) from the initial concentration $FeSO_4$ (90°C, $[PVA]_0$ = 2% mass., $[H_2O_2]_0$ = 1 mol/L).

3. Polyvinyl alcohol oxidation affected by hydrogen peroxide in the presence of ferrous sulfate (II) in aqueous solutions is accompanied by chemiluminescence in the visible spectrum (Fig. 23.6). In Fig. 23.6, it is seen that luminous intensity passes through a maximum. This facts shows that luminescence emitter is an intermediate product of liquid-phase PVA oxidation. The analysis of the kinetic curve of chemiluminescence intensity showed that its initial part is well linearized (Fig. 23.6, the correlation index $r = 0.998$) in the equation coordinates

$$I = g \times t^2,$$
(IV)

where g – a slope of the line.

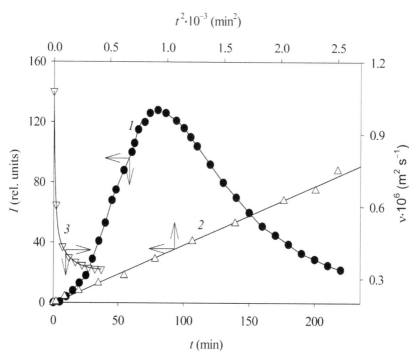

FIGURE 23.6   Kinetic curve of chemiluminescence signal intensity changes (*1*) and strengthening of its initial part in the equation coordinates (IV) (*2*); (kinetics of kinematical viscosity changes (*3*) (90°C, [PVA]$_0$ = 2% mass., [H$_2$O$_2$]$_0$ = 1 mol/L, [FeSO$_4$]$_0$ = 2×10^{-4} mol/L).

Comparing kinetics of kinematical viscosity reduction and chemiluminescence intensity changes (Fig. 23.6) shows that the intensity of luminescence variations inconsiderably changes during the initial period of rapid viscosity modifications. Thus, macromolecules of polyvinyl alcohol are first destructed and luminescence emitters are formed in the process of further oxidative transformations of the oligomers.

4. The kinetic scheme of the process. Analyzing of the received results and published data [16–21], the following scheme of radical oxidative transformations of polyvinyl alcohol including the stages of initiation, continuation and chain termination is offered.

5. Initiation. In the reaction system "PVA + H$_2$O$_2$ + O$_2$ + H$_2$O" the radical initiation takes place in disintegration of hydrogen peroxide (reaction (I)) catalyzed by impurities of variable valence metal ions [16–21].

In the system "PVA + O_3 + O_2 + H_2O" the initiation of the radical process is carried out according to the reaction of hydrogen-atom abstraction by ozone molecules from PVA C-H links leading to formation of secondary ($P^{\bullet}_{(1)}$) and tertiary ($P^{\bullet}_{(?)}$) alkyl radicals:

$$\sim CH_2CH^{\bullet}CHCHCH_2CH\sim \; + \; HO^{\bullet} + O_2 \quad (1.1)$$
$$\underset{OH}{|} \quad \underset{OH}{|} \quad \underset{OH}{|}$$
$$(P^{\bullet}_{(1)})$$

$$O_3 + \sim CH_2CHCH_2CHCH_2CH\sim$$
$$\underset{OH}{|} \quad \underset{OH}{|} \quad \underset{OH}{|}$$
$$(PH)$$

$$\sim CH_2^{\bullet}CCH_2CHCH_2CH\sim \; + \; HO^{\bullet} + O_2 \quad (1.2)$$
$$\underset{OH}{|} \; \underset{OH}{|} \quad \underset{OH}{|}$$
$$(P^{\bullet}_{(2)})$$

According to Refs. [22, 23], in aliphatic single and two-atomic alcohols a-CH-link of the substrate mainly reacts with the ozone. In our case tertiary PVA a-CH-links are apparently subjected to the ozone attack.

6. Continuation of the chain. The formed hydroxyl radicals on the initiation stage are quite active in the reactions of hydrogen atom separation from C-H links of oxygen-containing compounds. Thus, e.g. speed constants of $HO^{\bullet}$-radicals with mono- and polyatomic alcohols in an aqueous phase of 25°C are changed in the limits $(1.0 \div 5.0) \times 10^9$ L/mol·s [24, 25]. It is expected that hydroxyl radicals will display high activity in relation to all C-H links of polyvinyl alcohol:

$$\sim CH_2CH^{\bullet}CHCHCH_2CH\sim \; + \; H_2O \quad (2.1)$$
$$\underset{OH}{|} \quad \underset{OH}{|} \quad \underset{OH}{|}$$
$$(P^{\bullet}_{(1)})$$

$$HO^{\bullet} + \sim CH_2CHCH_2CHCH_2CH\sim$$
$$\underset{OH}{|} \quad \underset{OH}{|} \quad \underset{OH}{|}$$
$$(PH)$$

$$\sim CH_2^{\bullet}CCH_2CHCH_2CH\sim \; + \; H_2O \quad (2.2)$$
$$\underset{OH}{|} \; \underset{OH}{|} \quad \underset{OH}{|}$$
$$(P^{\bullet}_{(2)})$$

Let us estimate the possibility of forming secondary ($P^{\bullet}_{(1)}$) and tertiary ($P^{\bullet}_{(2)}$) radicals in the reactions (2.1) and (2.2) by the partial rate constants of $HO^{\bullet}$-radicals with C-H PVA bonds employing the data [24] on the reaction ability of the hydroxyl radical at 25°C in the reaction of H-atom separation from various alcohols. So using the speed constant of $HO^{\bullet}$-radicals with tetra-butanol ($5.3-10^8$ L/mol·s), $k_{\beta init} = (5.3-10^8)/9 = 5.89 \times 10^7$ L/mol·s was found (the subscript symbol points to the attacked CH-link). The partial speed constant of H-atom separation from the secondary a-CH-links was calculated from the speed constant value of reacting hydroxyl radical with ethylene glycol (1.6×10^9 L/mol·s) subject to four equal aCHlinks: $k_{a\text{-sec}} = (1.6-10^9)/4 = 4.0 \times 10^8$ L/mol·s. The amount of $k_{a\text{-tret}}$ was calculated using the value of constant speed reaction of HO˙-radical with 1.2 ($1.8-10^9$ L/mol·s) propandiol and the calculated values $k_{b\text{-init}}$ and $k_{a\text{-sec}}$:

$$k_{a\text{-tret}} = k_{HO˙ + propandiol - 1,2} - 3 \cdot k_{b\text{-init}} - 2 \cdot k_{a\text{-sec}} =$$
$$= 1.8 \times 10^9 - 3 \times 5.89 \times 10^7 - 2 \times 4.0 \times 10^8 = 8.2 \times 10^8 \text{ L/mol·s.}$$

And finally the partial speed constant of HO˙-radical reaction with bCHlink was found due to the reaction ability of hydroxyl radical in the reaction of 2-atom alcohols $HO(CH_2)_nOH$ (n = 2, 3, 4). As in the indicated row the speed reaction constant grows practically linear $1.6 \div 2.4 \div 3.1) \cdot 10^9$ L/mol·s, it may be considered that $k_{b\text{-sec}} \gg k_{a\text{-sec}} \gg 4.0 \times 10^8$ L/mol·s. From the data obtained $2 \cdot k_{b\text{-sec}} = 8.0 \times 10^8$ L/mol·s (2 CH-links account) and $k_{a\text{-tret}} = 8.2 \times 10^8$ L/mol·s it is followed that the possibility of forming $P'_{(1)}$ and $P'_{(2)}$ radicals in the reaction of HO˙ with PVA is quite equal (49% and 51% accordingly).

It is known through Refs. [16, 17, 19] that alkyl radicals rapidly join oxygen transforming to peroxyl radicals (with the speed constant $\sim 10^9$ L/mol·s):

$$\underset{(P'_{(1)})}{\sim CH_2CH˙CHCHCH_2CH\sim} + O_2 \longrightarrow \underset{(P_{(1)}O_2˙)}{\sim CH_2CHCHCHCH_2CH\sim} \quad (3.1)$$

$$\underset{(P'_{(2)})}{\sim CH_2˙CCH_2CHCH_2CH\sim} + O_2 \longrightarrow \underset{(P_{(2)}O_2˙)}{\sim CH_2CCH_2CHCH_2CH\sim} \quad (3.2)$$

So in the reaction systems "PVA + H_2O_2 + O_2 + H_2O" and "PVA + O_3 + O_2 + H_2O" characterized by quite large concentrations of the dissolved oxygen, the proportion of $P_{(1)}O_2^{\bullet}$ and $P_{(2)}O_2^{\bullet}$ peroxyl radicals will be considerably higher than the alkyl $P_{(1)}^{\bullet}$ and $P_{(2)}^{\bullet}$ ones.

In the work [19] it was shown that the oxidative destruction of carbon-chain polymers is connected with peroxyl and alkoxyl macroradicals. The oxidative destruction of polyvinyl alcohol evidently results in $P_{(1)}O_2^{\bullet}$ and $P_{(2)}O_2^{\bullet}$ radicals disintegration which are first subjected to isomerization:

$$
\begin{array}{c}
\underset{\underset{(P_{(1)}O_2^{\bullet})}{}}{\overset{\overset{OO^{\bullet}}{|}}{\sim CH_2\underset{|}{CH}\underset{|}{CH}CH_2CH\sim}} \longrightarrow \underset{}{\overset{\overset{OOH}{|}}{\sim CH_2\underset{|}{CH}\underset{|}{CH}CH^{\bullet}CHCH\sim}} \longrightarrow \\
OH \quad OH \quad OH \qquad\qquad OH \quad OH \quad OH
\end{array}
\tag{4.1}
$$

$$
\longrightarrow \underset{(r_{(1)}CHO)}{\sim CH_2\underset{\underset{OH}{|}}{CH}CHO} + HO^{\bullet} + \underset{OH \quad OH}{HC=CHCH\sim}
$$

$$
\begin{array}{c}
\underset{\underset{(P_{(2)}O_2^{\bullet})}{OH \quad OH \quad OH}}{\overset{\overset{OO^{\bullet}}{|}}{\sim CH_2\underset{|}{C}CH_2\underset{|}{C}HCH_2CH\sim}} \longrightarrow \underset{OH \quad OH \quad OH}{\overset{\overset{OOH}{|}}{\sim CH_2\underset{|}{C}CH_2^{\bullet}\underset{|}{C}CH_2CH\sim}} \longrightarrow
\end{array}
\tag{4.2}
$$

$$
\longrightarrow \underset{(r_{(3)}COOH)}{\sim CH_2C\overset{\nearrow O}{\underset{\searrow OH}{}}} + HO^{\bullet} + \underset{OH \quad OH}{CH_2=CCH_2CH\sim}
$$

Alongside with the reaction (4.2) the transformation leading to hydro-peroxyl radical extraction is considered quite efficient for the peroxyl radical $P_{(2)}O_2^{\bullet}$:

$$
\underset{\underset{(P_{(2)}O_2^{\bullet})}{OH \quad OH \quad OH}}{\overset{\overset{OO^{\bullet}}{|}}{\sim CH_2\underset{|}{C}CH_2\underset{|}{C}HCH_2CH\sim}} \longrightarrow \underset{\underset{(r_{(3)}C(O)r_{(5)})}{OH \quad OH}}{\sim CH_2\overset{\overset{O}{\parallel}}{C}CH_2CHCH_2CH\sim} + HO_2^{\bullet}.
\tag{4.3}
$$

The destruction of the carbon chain does not occur here.

The unrefined compounds formed in Eqs. (4.1) and (4.2) reactions are enols able to isomerize in the corresponding aldehydes and ketones:

$$HC=CHCH\sim \rightleftarrows \overset{O}{\underset{H}{C}}CH_2CH\sim ;$$
$$\underset{OH\ \ OH}{} \qquad \underset{OH}{}$$

$$(r_{(2)}CHO)$$

The products of oxidative destruction (aldehydes and ketones) will be subjected to further oxidative transformations. However, the reactivity of aldehydes in oxidative reactions is considerably higher as compared with ketones [16, 17]. On that assumption it is supposed that the final products of ozonized and peroxyl oxidation of polyvinyl alcohol (acids and CO_2) are mostly formed from oligomers of aldehyde groups.

7. Chain termination. The stages of radical process termination are mostly stages of recombining (disproportionation) of peroxyl radicals [16,17,19].

The scheme of PVA oxidative transformations can be generalized and presented in the following way:

$$H_2O_2 + M^{n+} HO^- + HO^{\bullet} + M^{(n+1)+} \tag{1}$$

$$O_3 + PH \left\langle \begin{array}{l} P^{\bullet}_{(1)} + HO^{\bullet} + O_2 \qquad\qquad (1.1) \\ P^{\bullet}_{(2)} + HO^{\bullet} + O_2 \qquad\qquad (1.2) \end{array} \right.$$

$$HO^{\bullet} + PH \left\langle \begin{array}{l} P^{\bullet}_{(1)} + H_2O \qquad\qquad (2.1) \\ P^{\bullet}_{(2)} + H_2O \qquad\qquad (2.2) \end{array} \right.$$

$$P^{\bullet}_{(1)} + O_2 \, P_{(1)}O_2^{\bullet} \tag{3.1}$$

$$P^{\bullet}_{(2)} + O_2 \, P_{(2)}O_2^{\bullet} \tag{3.2}$$

$$P_{(1)}O_2^{\bullet} \, r_{(1)}CHO + r_{(2)}CHO + HO^{\bullet} \tag{4.1}$$

$$P_{(2)}O_2^{\bullet} \left\langle \begin{array}{l} r_{(3)}COOH + CH_3C(O)r_{(4)} + HO^{\bullet} \qquad (4.2) \\ r_{(3)}C(O)r_{(5)} + HO_2^{\bullet} \qquad\qquad (4.3) \end{array} \right.$$

$$HO_2^{\bullet} + PH \quad\begin{cases} \longrightarrow P^{\bullet}_{(1)} + H_2O_2 & (5.1) \\ \longrightarrow P^{\bullet}_{(2)} + H_2O_2 & (5.2) \end{cases}$$

$$P_{(1)}O_2^{\bullet} + P_{(1)}O_2^{\bullet}\, \Pi_{6.1} \tag{6.1}$$

$$P_{(2)}O_2^{\bullet} + P_{(2)}O_2^{\bullet}\, \Pi_{6.2} \tag{6.2}$$

$$rCHO + HO^{\bullet}\ rC^{\bullet}{=}O + H_2O \tag{7}$$

$$rC^{\bullet}{=}O + O_2\ rC(O)OO^{\bullet} \tag{8}$$

$$rC(O)OO^{\bullet} + rCHO\ rC(O)OOH + rC^{\bullet}{=}O \tag{9}$$

$$rC(O)OO^{\bullet} + rC(O)OO^{\bullet}\, \Pi_{10} \tag{10}$$

$$rC(O)OOH + M^{n+}\ rC(O)O^{\bullet} + HO^{-} + M^{(n+1)+} \tag{11}$$

$$rC(O)O^{\bullet} + rCHO\ rCOOH + rC^{\bullet}{=}O \tag{12}$$

$$rC(O)O^{\bullet}\ r^{\bullet} + CO_2 \tag{13}$$

$$r^{\bullet} + O_2\ rO_2^{\bullet} \tag{14}$$

$$rO_2^{\bullet} + rO_2^{\bullet}\, \Pi_{15} \tag{15}$$

In the given scheme $\Pi_{6.1}$, $\Pi_{6.2}$, Π_{10} and Π_{15} are products of destruction of peroxyl radicals (6.1) (6.2) (10) and (15), respectively; $r_{(1)}$, $r_{(2)}$, $r_{(3)}$, $r_{(4)}$, $r_{(5)}$, r are fragments of PVA oligomers with different number of structural elements.

The analysis of experimental data shows that the oxidation of polyvinyl alcohol under hydrogen peroxide and ozone-oxygen compounds in an aqueous medium can be divided into two processes for convenience: oxidative destruction of PVA macromolecules and oxidative functionalization of the formed oligomers.

23.3.1 OXIDATIVE DESTRUCTION OF POLYVINYL ALCOHOL MACROMOLECULES

The limiting stage of destruction is probably a stage of peroxyl macroradical disintegration ($P_{(1)}O_2^{\bullet}$ or $P_{(2)}O_2^{\bullet}$). In this case oligomers containing aldehyde, carbonyl and carboxyl groups (reactions (4.1)–(4.3)) are formed.

According to the scheme, in the stationary mode the speed of oligomer formation with the aldehyde group is equal to:

$$d[rCHO]/dt = k_{4.1} / 2\sqrt{2k_{6.1}} \cdot \sqrt{V_i} \qquad (V)$$

Integration (V) (assuming that V_i = const) leads to the following equation of the kinetic curve of aldehyde accumulation:

$$[rCHO] = (k_{4.1}/2\sqrt{2k_{6.1}} \cdot \sqrt{Vi}) \cdot t = d \cdot t. \qquad (VI)$$

Similarly, the equation for the speed of acid accumulation can be obtained on applying the condition of quasi-stationary to $P_{(2)}O_2^\cdot$ radical.

$$[rCOOH] = (k_{4.2}/2\sqrt{2k_{6.2}} \cdot \sqrt{Vi}) \cdot t = a \cdot t. \qquad (VII)$$

Thus, at destruction of $P_{(1)}O_2^\cdot$ and $P_{(2)}O_2^\cdot$ radicals aldehydes and acids are with time accumulated due to the same linear law (compare (VI) and (VII) equations).

23.3.2 OXIDATIVE FUNCTIONALIZATION OF OLIGOMERS

Under the scheme the formed aldehydes are oxidized further to peroxy acids with the speed:

$$d[rCOOH]/dt = ((k_9 \cdot d \cdot \sqrt{V_i})/(\sqrt{2k_{10}})) \cdot t \qquad (VIII)$$

where the following equation of the kinetic curve of peroxy acid accumulation is given:

$$[rC(O)OOH] = ((k_9 \cdot d \cdot \sqrt{V_i})/(2\sqrt{2k_{10}})) \cdot t^2. \qquad (IX)$$

The final products of oxidative transformations of polyvinyl alcohol (CO_2 and acids) are probably formed in non-limiting stages (12) and (13). On applying the condition of quasi-stationary to $rC(O)O^\cdot$ radical there can be given the following equations of kinetic curves of acid and carbon dioxide accumulation:

$$[rCOOH] = ((a \cdot k_9 \cdot d \cdot \sqrt{V_i})/(2\sqrt{2k_{10}})) \cdot t^2 = b \cdot t^2 \qquad (X)$$

$$[CO_2]=((\beta \cdot k_9 \cdot d \cdot \sqrt{V_i})/(2\sqrt{2k_{10}})) \cdot t^2 = c \cdot t^2 \qquad \text{(XI)}$$

where a and b are proportions of peroxy acids leading to forming acidic reaction products and CO_2 accordingly.

From the expressions (X) and (XI) it is seen that acids and carbon dioxide must be in time accumulated due to the same law. However, for rCOOH there exists one more formation channel, namely the oxidative destruction of radicals $P_{(2)}O_2^{\cdot}$ in the Eq. (4.2). Then taking into account the formulas (VII) and (X) there can be obtained the final equation of the kinetic curve of acid accumulation:

$$[rCOOH] = a \times t + b \times t^2, \qquad \text{(XII)}$$

which coincides with the empirical dependence (I) (Fig. 23.1).

Thus, the above scheme of oxidative transformations of polyvinyl alcohol allows to explain the received experimental results. Indeed, the main part of acidic products (see the equations (I) and (XII)) and carbon dioxide (see the equations (II) and (XI)) are formed by the quadratic law in the process of radical oxidation of aldehydes. The inconsiderate channel of acid accumulation is determined by $P_{(2)}O_2^{\cdot}$ radical destruction.

The data of the reaction system "PVA+ H_2O_2 + O_2 + H_2O" is used for estimating the relation of rCOOH and CO_2 accumulation speeds. Comparison of b and c parameters shows that under comparable conditions (90°C, $[PVA]_0 = 2\%$ mass., $[H_2O_2]_0 = 1$ mol/L) the amount of acids is more than carbon dioxide:

b/c= $(2.77 \times 10^{-7} \cdot \text{mol}/(\text{L} \cdot \text{min}^2))/((1.73 \times 10^{-7} \cdot \text{mol})/(\text{L} \cdot \text{min}^2))=1.6$

The analysis of the kinetic equations curves definitely indicates that accumulation of PVA oxidation products considerably depends on the initiation speed (V_i). Here it is the stability of V_i rather than the way of radical generation that is important (H_2O_2 disintegration, ozone-oxygen mixture bubbling or application of other initiators). This is probably explained by the same kinetic laws of polyvinyl alcohol oxidative transformations in the presence of different initiating systems (H_2O_2, $O_3 + O_2$).

The chemiluminescent data (Fig. 23.6) received in the reaction system "PVA + H_2O_2 + O_2 + H_2O" in introducing ferrous sulfate (II) into it may serve an additional confirmation of the above suggested scheme of PVA radical oxidative transformations.

It is known that in reactions of oxidizing organic substances by oxygen [18, 26, 27] and ozone-oxygen mixture [28–30] chemiluminescence occurs in element acts of disproportionation of peroxyl radicals. The products of these element acts are electronically excited carbonyl compounds. It may be expected that in polyvinyl alcohol oxidation under H_2O_2 and $FeSO_4$ presence, the excited carbonyl compounds $>C=O^*$ formed in the strongly exothermic act of disproportionation of peroxyl radicals of PVA oligomers will serve as luminescent emitters (see the kinetic scheme of the process, reaction (15)):

$$rO_2^{\cdot} + rO_2^{\cdot} (\eta_1) > C=O^* + (1-\eta_1) > C=O + products,$$
$$>C=O^* >C=O + hn,$$

where η_1 and η_2 – outputs of excitation and emission $>C=O^*$, respectively, $h = \eta_1 \times \eta_2$ – chemiluminescence output.

This approach allows to explain the quadratic time dependence of chemiluminescence intensity changes (Eq. (IV)) describing the initial part of the kinetic curve (Fig. 23.6). Indeed, the chemiluminescence intensity is connected with emitter concentration by the relation:

$$I = \eta_2 \cdot k_{16}[>C=O^*] = \eta_1 \cdot \eta_2 \cdot k_{15}[rO_2^{\cdot}]^2 = \eta \cdot k_{15}[rO_2^{\cdot}]^2. \tag{XIII}$$

Applying the quasistationary conditions to the radicals $rO_2^{\cdot}$, $r^{\cdot}$ and $rC(O)O^{\cdot}$ the following expression is given:

$$I = 2h \cdot \beta \cdot k_{11}[M^{n+}][rC(O)OOH]. \tag{XIV}$$

As $[rC(O)OOH]$ is proportional to t^2 (Eq. (IX)), substituting (IX) into (XIV) the final formula for chemiluminescence intensity is received:

$$I = ((\eta \cdot \beta \cdot k_9 \cdot k_{11} \cdot d \cdot [M^{n+}] \cdot \sqrt{V_1}) / (\sqrt{2k_{10}})) \cdot t^2 = g \cdot t^2 \tag{XV}$$

which coincides with the empiric dependence (IV) (Fig. 23.6).

Diminishing of chemiluminescence intensity on the profound stages is caused by several factors: decreasing of the initiating speed in the reaction (I) due to reducing the concentration of hydrogen peroxide; extinguishing of the emitters by oxidation products; increasing of speed of nonradical acid-catalytic disintegration of peroxides as far as the acid is accumulated in the system.

Thereby, the received kinetic laws of chemiluminescence intensity changes may be considered as an additional confirmation of the suggested scheme of radical oxidative PVA transformations. The same quadratic laws of acid accumulation (Eqs. (I) and (XII), CO_2 (Eqs. (II) and (XI)) and chemiluminescent emitters (Eqs. (IV) and (XV)) testify that all of them are formed in the process of oligomer oxidation with aldehyde groups which in their turn are products of oxidative destruction of polyvinyl alcohol macromolecules.

KEYWORDS

- **Destruction**
- **Functionalization**
- **Hydrogen Peroxide**
- **Kinetics**
- **Oxidation**
- **Oxygen**
- **Ozone**
- **Polyvinyl Alcohol**
- **Radical Mechanism**

REFERENCES

1. Ushakov, S. N. (1960). Poliviniloviy spirt i ego proizvodnye. Moscow, 552p (in Russian)
2. Sakurada, I., & Matsuzawa, S. (1959). Chem. High Polym. 16, 633.
3. Sakurada, I., & Matsuzawa, S. (1961). Chem. High Polym. 18, 252.
4. Vink, H. (1963). Makromolek. Chem. 67, 105.
5. Trudelle, Y., & Neel, J. (1964). C.R. Acad. sci., 258, 4267.
6. Trudelle, Y., Neel, J. (1964). C.R. Acad. sci. 258, 4542.
7. Schlapfer, K. (1965). Schweiz. Arch. angew. Wiss. und Techn. 31, 154.
8. Mino, G., Kaizerman, S., & Rasmussen, E. (1959). J. Polymer Sci. 39, 523.
9. Ikada, Y., Nishizakii, Y., Iwata, H., & Sakurada, I. (1977). J. Polym. Sci., 15, 451.
10. Staszewska, P., Ulinska, A., & Pomagala, B. (1973). Polimerytworz. Wielkoczasteczk. 18, 459.
11. Sakurada, I., & Matsuzawa, S. (1963). Chem. High Polym. 20, 349.
12. Ibonai, M. (1964). Polymer. 5, 317.
13 Vendillo, V. P., Yu, M., & Emelyanov, Yu. V. (1959). Philippov, Zavodskaya laboratoriya, 25, 1401 (in Russian).

14. Yu, S., Zimin, A. F., Ageeva, K. G., Serebrennikova, I. M., & Borisov, Yu. B. (2004). Monakov, Izv. Vuzov. Khimiya I Khim. Tekhnologiya, 47, 37(in Russian).
15. Yu, S., Zimin, A. F., Ageeva, A. V., Yanysheva, I., Borisov, M., & Yu, B. (2004). Monakov, Izv. Vuzov. Khimiya I., Khim. Tekhnologiya, 47, 119(in Russian).
16. Emanuel, N. M., Zaikov, G. E., & Maizus, Z. K. (1973). Rol sredy v radicalno-cepnykh reakciyakh okisleniya organicheskikh soedineniy. Nauka, M 279p (in Russian).
17. Denisov, E. T., Mitskevitch, N. I., & Agabekov, V. E. (1975). Mekhanism zhidkop-haznogo okisleniya kislorodsoderzhaschikh soedineniy. Minsk, Nauka i tekhnika 334p (in Russian).
18. Denisov, E. T. (1981). Mekhanizmy gomolititcheskogo raspada molekul v zhidkoy faze Itogi nauki I tekhniki. Seriya "Kinetika I kataliz" T, 9–M, VINITI, 158p (in Russian).
19. Denisov, E. T. (1990). Okisleniye I destruktsiya karbotsepnikh polimerov. L., Khimiya, 288 p (in Russian).
20. Sytchev, A., Ya., & Isaak, V. G. (1995). Usp. Khimii 64, 1183(in Russian).
21. Antonovskiy, V. L., & Khursan, S. L. (2003) Fisitcheskaya khimiya organitcheskikh peroksidov M., Akademkniga, 391 p (in Russian).
22. Shereshovets, V. V., Ya, N., & Shafikov, V. D. (1980). Komissarov, Kinetika i Kataliz, 21, 1596 (in Russian).
23. Komissarov, V. D., Yu, S., Zimin, N. V., & Trukhanova, & Zaikov, G. E. (2005). Oxid. Commun, 28, 559.
24. Pikaev, A. K., & Kabakchi, S. A. (1982). Reaktsionnaya Sposobnost Pervichnykh Produktov Radioliza vody, Spravotchnik, M. Enegroizdat 200p (in Russian).
25. Denisov, E. T., & Denisova, T. G. (1994). Izv. A. N. Ser. khim. 1, 38 (in Russian).
26. Ya, V., Shlyapintokh, O. N., Karpukhin, L. M., Postnikov, I. V., Zakharov, A. A., & Vitchutinskiy, V. F. (1966). Tsepalov, Khemiluministsentnye Metody Issledovaniya Medlennikh Khimicheskikh Protsessov. Nauka M., 300p (in Russian).
27. Vasiliev, R. F. (1966). Usp. phis. nauk. 89, 409 (in Russian).
28. Gusmanov, A. A., Agapitova, N. A., Yu. S., & Zimin, I. M. (2004). Borisov, React. Kinet. Catal. Lett. 81, 357.
29. Yu, S., Zimin, A. A., & Gusmanov, S. L. (2004). Khursan, Kinetika I kataliz, 45, 829 (in Russian).
30. Yu, S., Zimin, Yu. B., Monakov, V., & Komissarov, D. (2005). Chemical and biological kinetics. New horizons. V. 1:Chemical kinetics Burlakova, E. B., Shilov, A. E., Varfolomeev, S. D., Zaikov, G. E., (eds). Brill Academic Publishers, Leiden, Boston, Inc. 172–188.

CHAPTER 24

ACTIVITY OF HYDROLYTIC ENZYMES OF BASIDIOMYCETES ACCORDING TO THEIR TROPHIC FEATURES

I. A. SHPIRNAYA, V. O. TSVETKOV, Z. A. BEREZHNEVA, and R. I. IBRAGIMOV

CONTENTS

ABSTRACT

In the samples of fruit bodies of basidiomycetes level of activity of the proteolytic, cellulolytic and pectolytic enzymes is very high. Xylotrophic mushrooms obtain a high amylolytic activity. In saprotrophic fungi the amylases activity is absent. Mycorrhizal saprotrophs have a high pectinase activity as compared with the Litter saprotrophs

24.1 INTRODUCTION

Community of fungi is one of the main functional and structural components of forest ecosystems. Their species content is quite diversified as species have different functions in different trophic levels.

Fungi are fed via osmosis, generally (parasites, symbionts and saprotrophes) with plants tissues. A fungal feeding feature is their ability of synthesis of extracellular enzymes. Fungal organisms decompose such complex biopolymers as cellulose, lignin, pectin substances, hemicellulose, proteins and form an important step of biological degradation and re-synthesis of organic substances in nature [1]. Fungal enzymes are different as regarding both their source and activity. Hydrolytic enzymes, degrading polymers of plant cell, obtain a special role in this complex. They include carbohydrases, degrading some olygo- and polysaccharides, proteolytic enzymes and esterases, degrading phosphorus and ester bonds, etc.

One of the most important ways of development of modern microbiology and biotechnology is technologies of producing some biologically active substances, including enzymes. Now enzyme agents that are obtained via microscopic fungi cultivating are widely used. However, there are problems in their usage as products of some substances [2].

Higher fungi can be used as a source of enzymes both for feeding and medical usage. Advantages of these fungi in artificial cultivating are the ability for feeding and absence of sporulation in culture, so any danger of job diseases is low [2, 3].

Therefore, the research of enzyme activity on higher fungi is topical for both science and practice.

The purpose of our research is to determine the activity level of hydrolytic enzymes (proteases, cellulases, pectinases and amylases) in fruiting bodies of higher basidial fungi of different trophic groups.

24.2   METHODS

The objects of our research were fruiting bodies of wild species of some families of basidiomycetes, harvested at autumn (September–October) in the area of Ufa in Republic of Bashkortostan (Table 24.1). Growing, medium-sized and thick consistent mushrooms without wormholes were selected for research and were freezed at −20°C closely after harvesting.

TABLE 24.1   Taxonomy of Fungi – Basifiomycetes

No.	Latin name	Family
1	Suillus luteus	Suillaceae
2	Lactarius deliciosus	
3	Russuia rubra	Russulaceae
4	Lactarius resimus	
5	Clitocybe geotropa	
6	Clitocybe nebularis	
7	Ganoderma lucidum	Ganodermaceae
8	Agaricus arvensis	Agariceae
9	Amanita muscaria	Amanitaceae
10	Leccinum scabrum	Boletaceae
11	Armillariella mellea	Tricholomataceae

For producing extracts with the enzyme activity the freezed fruit bodies were homogenized in porcelain pounder with quartz sand and extracted with distilled water in the ratio 1:1. Next steps were incubating in 1.5 h at 4°C, filtering and double centrifugation at 10,000 rpm (11,000 g) (centrifuge MPW-310, Poland) for 10 min.

To evaluate the concentration of water-soluble protein, the Bradford method was used. The calibrating plot for evaluation of protein concentration was generated using alpha-chymotrypsin.

For measuring the activity of hydrolytic enzymes the method of agarose gel plates was applied [4]. For measuring the proteases activity a gelatin was used as a substrate, for amylases – starch, for cellulases – carboxymethylcellulose (CMC), for pectinases the apple pectin was used. The concentration of substrate in gel was always 1%. To make the reaction of substrate hydrolysis reaction the plates with the samples were incubated in a wet camera for 18 h at 25 °C. To visualize the areas of hydrolyzed substrate on the plates the authors used the following reagents: for proteinases – 5% acetate, for amylases – Lugol's solution, for cellulases – 6% lead acetate (CMC), for pectinases – 10% copper acetate (pectin). The activity of enzymes was evaluated via measuring of the gel area with hydrolyzed substrate around the hole. The enzyme quantity that hydrolyzed the substrate in 1 mm^2 area was used as 1 milliunit of the enzyme activity (mE).

Experiments were carried out with three analytic and three biological repeats.

24.3   RESULTS AND DISCUSSION

More than one hundred of enzyme preparations were extracted, purified and characterized from basidiomycetes, however, searching of novel species of fingi is still topical [3]. Feed features of fungi are associated with their specific (via osmosis) feeding way and contents of the substrate they live. These facts are one of the factors determining the contents of secreting enzyme complex [6].

The activity of hydrolytic enzymes, playing the main role in degradation of fungi food substrates (proteases, carbohydrases), was researched. Our research shows (Table 24.2), that all the samples obtain high proteolytic activity. This is probably due to the fact that the protease is performed in a variety of physiological functions of the living body, from the digestion of food substrate proteins to specific regulatory processes such as the activation of zymogens, formation of hormones and other physiologically active peptides from their precursors, proteins transport, protective responses, etc. [7, 8]. Most of the proteolytic activity of extracts of fruit bodies had *Suillus luteus* and *Leccinum scabrum*, which amounted to 9.74 and 11.01 E/g, respectively. The extract of *Lactarius resimus* and *Agaricus arvensis* had the least activity of proteases among the studied samples.

TABLE 24.2 The Activity of Extractive Hydrolytic Enzymes of Basidiomycetes Fruit Bodies, E/g of Wet Weight

Sample	Cellulose	Amylase	Protease	Pectinase	Water-soluble protein, mg per g
Suillus luteus	3.17±0.49	0	9.7±1.02	4.96±0.38	1.75 ±0.009
Lactarius deliciosus	5.19±0.44	0	7.42±0.45	4.53±0.35	2.01 ±0.009
Russuia rubra	4.32±0.40	0	6.60±0.71	6.13±0.48	2.20±0.017
Lactarius resimus	3.92±0.54	0	3.54±0.37	4.74±0.60	3.05±0.009
Clitocybe geotropa	6.13±0.48	0	7.97±0.90	3.73±0.32	1.50±0.009
Clitocybe nebularis	4.50±0.35	0	3.35±0.58	4.10±0.35	1.45±0.017
Ganoderma lucidum	3.54±0.37	3.54±0.82	8.25±0.56	4.12±0.64	3.05±0.007
Agaricus arvensis	6.13±0.87	0	2.83±0.33	3.10±0.49	3.05±0.009
Amanita muscaria	4.70±0.60	0	6.65±0.71	4.12±0.35	4.02±0.009
Leccinum scabrum	3.93±0.54	0	11.01±0.55	5.41±0.38	1.50±0.009
Armillariella mollea	3.53+0.47	0	5.71±0.53	4.36±0.53	1.35±0.009

Cellulases actively destroy the substrate on which the mushrooms grow in the wild. In general, the cellulolytic activity is characteristic of all the studied samples, the highest activity was detected in the fruit bodies of *Agaricus arvensis* and *Clitocybe geotropa*. Extracts of *Suillus luteus*, *Lactarius resimus* and *Leccinum scabrum* had the least activity of these enzymes (Tables 24.2 and 24.3).

TABLE 24.3   The Activity of Amylolitic Enzymes in Different Mushrooms

Sample	Activity, E/g of wet weight
Laetiporus sulfureus	127.67±1.31
Piptoporus betulinus	82.52±0.65
Fomitopsis officinalis	41.28±1.13
Trametes versicolor	74.66±1.13
Fomitopsis pinicola	44.18±0.65
Fomes fomentarius	220.35±1.13

As it was found, the highest pectinase activity had fruiting bodies of *Russuia rubra* and *Leccinum scabrum*, and the least – *Agaricus arvensis* and *Clitocybe*. Thus, fungi related to environmental group of mycorrhizal saprotroph (*Russuia, Leccinum*) when compared with the group of Litter saprotroph (*Agaricus, Clitocybe*) has higher pectinase activity, which is consistent with the literature [1]. The amylolytic activity was detected only in *Ganoderma* extracts. There is evidence on the constitutive nature of biosynthesis of amylase in wood-destroying fungi [1, 9], which is probably due to their trophic characteristics.

The remaining samples did not detect this activity, which may be related to the phenomenon of substrate induction. This phenomenon has been described for the amylolytic enzymes [10].

It is known that the biological value of the proteins of microbial and fungal biomass may exceed the value of the proteins of cereals and legumes. A promising direction for producing of a protein is to use fungal biomass for this purpose [5]. The authors determined the dependence of the water-soluble protein content on the species of the studied fungi (Table 24.2).

In general, all the samples investigated obtain a sufficiently high content of the soluble protein. As it can be seen, the greatest amount of protein per gram of wet weight was contained in the fruiting bodies of *Amanita* and *Ganoderma*, which may be due to the presence of a smaller amount of water in their fruiting bodies. Also, high protein concentration was observed in fruiting bodies of *Agaricus* and *Lactarius*. These results confirm the data of mushrooms as a valuable source of dietary protein.

The presence of amylolytic activity only in extracts of *Ganoderma* causes interest in the study of xylotrophic fungi. Therefore, several samples of *Ganoderma* were selected, where the amylolytic activity was determined.

There are about 100 species of xylotrophic fungi in Basidiomycota Division in the forests of the Southern Urals. Basidial macromycetesxylotrophes are one of the most important trophic groups of fungi in the forest ecosystems. Xylotrophes fungi make a stepwise decomposition of lignin and cellulose wood complexes, contribute to the formation of humus and transform various microelements into the food chain. Some species have dietary fruit bodies and are harvested for food purposes. In recent years wood-destroying fungi increasingly attracts the attention of pharmacologists as producers of biologically active substances having antioxidant and adaptogenic properties. Among these species there is a common group of *Ganoderma* belonging to the species of wood-destroying fungi of *Basidiomycetes* class (*Aphyllophorales* order).

The following species were researched: *Laetiporus sulfureus, Piptoporus betulinus, Fomitopsis officinalis, Trametes versicolor, Fomitopsis pinicola, Fomes fomentarius.*

As you can see, all of the samples display a significant amylase activity. The highest rates of activity of amylolytic enzymes are characteristic of *Laetiporus sulfureus* and *Fomes fomentarius,* it is 127.67 and 220.35 E/g of wet weight, respectively.

24.4   CONCLUSION

Thus, in the samples of fruit bodies of basidiomycetes level of activity of the proteolytic, cellulolytic and pectolytic enzymes is very high. Xylotrophic mushrooms obtain a high amylolytic activity. In saprotrophic fungi the amylases activity is absent. Mycorrhizal saprotrophs have a high pectinase activity as compared with the Litter saprotrophs.

KEYWORDS

- Amylase
- Basidiomyces
- Cellulase
- Hydrolytic Enzyme
- Pectinase
- Protease
- Protein

REFERENCES

1. Daniljak, N. I., Semichaevskij, V. D., Dudchenko, L. G., & Trutneva, I. A. (1989). Fermentnye Sistemy Vysshih Bazidial'nyh Gribov-Kiev, Nauk, Dumka, 280p.
2. Petrov, P. T., Skripko, A. D., & Litvinova, K. V. (2006). Novye Lekarstvennye Sredstva Na Osnove Biologicheski Aktivnyh Soedinenij Micelial'nyh gribov, Uspehi medicinskoj mikologii, 7, 198–199.
3. Psurceva, N. V., Kijashko, A. A., & Shahova, N. V. (2007). Jekologo-Taksonomicheskie Predposylki Poluchenija Plodovyh Tel v kul'ture Makromicetov, Predstavljajushhih Interes dlja mediciny, Uspehi Medicinskoj Mikologii, T. IX, 254–258.
4. Shpirnaja, I. A., Umarov, I. A., Shevchenko, N. D., & Ibragimov, R. I. (2009). Opredelenie Aktivnosti Gidrolaz i ih Ingibitorov Po gidrolizu Substrata v Gele Agarozy, Prikladnaja Biohimija i Mikrobiologija, 45(4), 497–501.
5. Ufimceva, O. V. (2006). Harakteristika Aminokislotnogo Sostava Glubinnoj Kul'tury Gribov Veshenki Obyknovennoj (Pleurotus ostreatus) i Serno-zheltogo Trutovika (Laetiporus sulphureus), Vestnik Biotehnologii Fiziko-himicheskoj Biologii Imeni Ju, A., Ovchinnikova, 2(4), 52–54.
6. D'jakov Ju, T. (1997). Griby i ih Znachenie v Zhizni Prirody i Cheloveka, Sorosovskij Obrazovatel'nyj Zhurnal, 3, 38–45.
7. Buhalo, A. p. (1983). Vysshie Bazidiomicety Producenty Pishhevogo Belka v Glubinnoj Kul'ture, Tez. dokd. vsesojuzn.konf. i micelial'nye griby, Pushhino, 99p.
8. Mosolov, V. V. (1971). Proteoliticheskie Fermenty, Nauka, M., 414p.
9. Bekker, Z. (1988). Je Fiziologija i biohimija gribov, M., izd-vo Mosk. un-ta, 230p.
10. Metlickij, L. V., & Ozereckovskaja, O. L. (1985). Kak rastenija Zashhishhajutsja ot boleznej. Nauka, M., 192p.

INDEX

Milton Keynes UK
Ingram Content Group UK Ltd.
UKHW022102141024
449569UK00031B/1740